CLIMATE CHANGE

CLIMATE

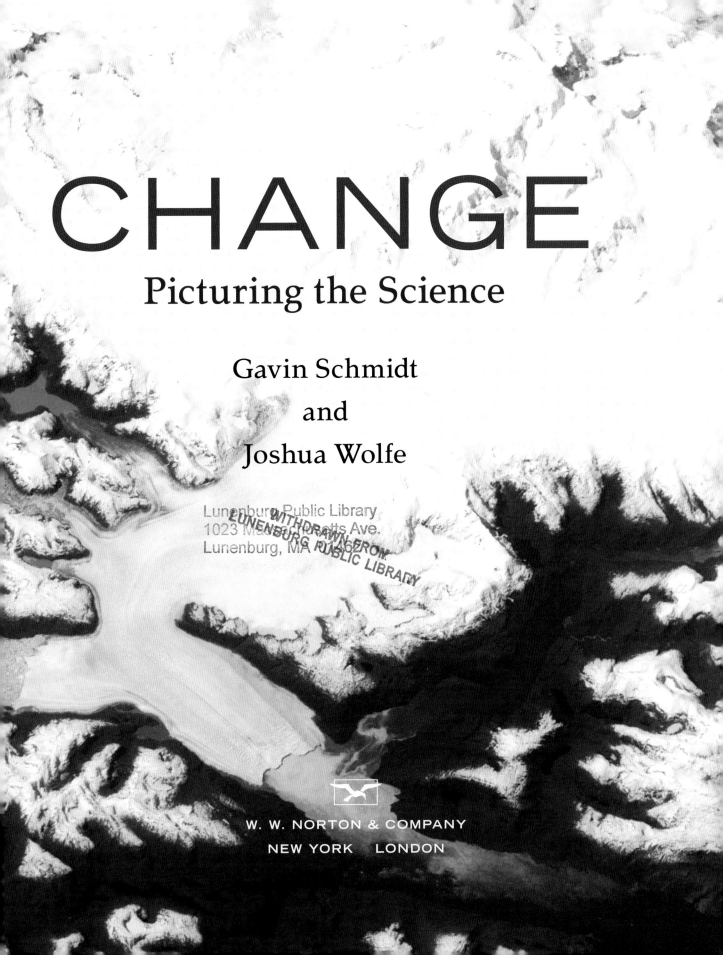

CHANGE
Picturing the Science

Gavin Schmidt

and

Joshua Wolfe

W. W. NORTON & COMPANY

NEW YORK   LONDON

CONTENTS

12/12

FOREWORD

Heading off the worst of human-induced climate change will require global cooperation. No single country or region can solve the climate crisis on its own. Moreover, governments around the world will act only when their citizens understand the magnitude of the challenge, the risks of continuing with business as usual, and the options for action. An informed public, therefore, is essential for the world to find effective solutions for one of the most harrowing and complex challenges facing humanity. Yet with a challenge so complex, so encompassing, and with so much inherent uncertainty, finding a path to public understanding and responsible action is a vast challenge in its own right.

Climate Change: Picturing the Science is a tour de force of public education. It is simply the best available collection of essays by climate scientists on the nature of human-induced climate change, the ways scientists have come to understand and measure the risks that it poses, and the options that we face. I am, of course, hugely proud that it is largely (though by no means wholly) the work of scientists at Columbia University's Earth Institute, a unique cross-faculty initiative by the university to bring science to bear on the global challenges of sustainable development. This book is an exemplar of what public education in the twenty-first century can and should accomplish.

The editors, climatologist Gavin Schmidt and photographer Joshua Wolfe, have produced a collection of essays of uniformly outstanding quality, supported by photographs of beauty and insight. Each chapter offers a scientifically rich, yet remarkably jargon-free, account of one crucial aspect of the climate change challenge. Several essays, including the one by Peter deMenocal and Tim Hall and the one by Gavin Schmidt, describe the underlying human and natural processes that lead to human-induced climate change, explaining the direct effects of greenhouse gases and the ways these effects may be amplified by various positive feedbacks in the climate system. These and the accompanying essays that describe how climatologists measure and verify climate change are told vividly by scientists who have been at the very forefront of these challenges.

Several powerful essays explain why human-induced climate change matters, and matters urgently. A scintillating essay by Shahid Naeem describes how climate change is already impacting the entire biosphere—the thin film of life covering the Earth—with a remarkably complex range of effects that can threaten the basic functioning of ecosystems and deprive them of their resilience and productivity and their ability to provide services—such as food, fresh water, and a safe environment—to humanity. The essay also makes clear the pervasive threats to the very survival of millions of other species as well. Another beautifully written chapter by Adam Sobel takes on the complex issue of extreme weather events, describing how and why climate change is likely to increase the frequency of droughts, floods, heat waves, and high-intensity tropical cyclones in some parts of the world. An essay by Stephanie Pfirman documents the changes already under way in the Arctic and Antarctica, with ramifications that will threaten human societies and biodiversity far beyond the Arctic itself. In all of these chapters, powerful photographs help to illuminate a complex and compelling story.

After these careful, precise, and yet highly accessible accounts of the underlying science, the collection then turns to the choices facing humanity. What do we know about the prospects for future climate change if we stay on the current business-as-usual trajectory, or if humanity adopts an alternative path based on new technologies for energy use, agriculture, and patterns of urbanization and land use? What are the likely costs and benefits of alternative public policy choices? What time horizons are involved? And what can one individual do to contribute to global solutions?

The essayists avoid glib solutions and stay true to the science, with all of its uncertainties and incomplete answers. Yes, we must make choices, but no, we cannot know with utter precision the costs and benefits (and for whom) of these alternatives. Gavin Schmidt explains clearly the hows and whys of climate scenarios, which summarize the likely outcomes of alternative policy choices. Frank Zeman explains how new ways of producing and deploying commercial energy—by replacing fossil fuels with renewable sources, by economizing on energy use through improved technologies for automobiles and buildings, or by capturing and safely storing the carbon dioxide emitted by fossil fuel power plants and factories—can dramatically reduce the buildup of greenhouse gases in the atmosphere, thereby reducing the human impact on climate change. A concluding essay by David Leonard Downie, Lyndon Valicenti, and Gavin Schmidt describes how this massive challenge must, in the end, be addressed by an equally massive effort at global problem solving, a "preventative planetary care" requiring an unprecedented level of global cooperation. The essay brings us up to date with the global efforts to reach consensus within the framework of international law, notably the UN Framework Convention on Climate Change, and the accompanying globally agreed protocols needed to implement the convention.

There are no shortcuts to addressing a challenge that is global, pervasive, profound, and long term. Global citizens must grasp the challenge, master its intricacies, and take responsibility, for our own generation and those to come. Recent years have seen a dramatic increase in global awareness and concern. *Climate Change: Picturing the Science* provides another vital impetus for understanding and therefore for action. For this reason, the book elicits enormous admiration for the science of climatology as well as enormous hope that humanity, based on a deeper awareness of the science and public policy of climate change, will rise to the challenge of protecting human well-being and the beauty and rich diversity of the Earth.

Jeffrey D. Sachs
New York
June 16, 2008

PREFACE

In interactions on the Web or at public talks and exhibitions, we have found a hunger among the public for more context and information about climate change that is not being satisfied by newspapers, television, or the occasional documentary. Elsewhere, photographers have begun documenting the effects of current climate change and have created images that bring home the depth and immediacy of the problem. Stemming from a 2005 gallery exhibition, "Photographers' Perspectives on Global Warming," in New York, this book is a marriage (it is hoped a happy one) between the image makers and the investigators. We have selected images on their intrinsic merit, and we have used the text to provide the background necessary to understand what is being seen and discussed. This is not a textbook, nor just a collection of pretty pictures. Instead, our aim is to provide an accessible summary of the state of the science and a visual record of what it means.

Interest in human-induced climate change as a public issue has a long and varied history. It goes back at least to Swedish chemist Svante Arrhenius's first calculation of the effect of increasing greenhouse gases on temperature in the late nineteenth century. It was recognized as an environmental problem by President Johnson as early as 1965. NASA scientist James Hansen's testimony to Congress in the summer of 1988 made newspaper headlines, while the Rio Earth Summit in 1992 and the 2006 release of Al Gore's film documentary *An Inconvenient Truth* have heightened public concerns. In previous decades, the issue rose in the public consciousness temporarily, only to be subsumed as more immediate concerns vied for media attention. Scientifically, however, the study of climate change has proceeded at a steady pace, and evidence of human modification of the climate has been mounting rapidly.

With the 2007 publication of the Fourth Assessment Report of the Intergovernmental Panel on Climate Change, a group of scientists sponsored by the United Nations and the World Meteorological Organization, the case for global warming has become "unequivocal," with a "very likely" dominant role for humans causing it.

But what lies behind these definitive statements? Where are the key observations and theoretical insights that climate scientists rely on? Are there remaining issues? What does climate change portend? These are questions now asked daily at the water cooler, in the newsroom, and in Congress.

The scope of climate change is truly vast, and no one author can do justice to its varied aspects. Accordingly, we have brought together experts on atmospheric science, oceanography, paleoclimatology, the polar regions, technology, and politics to each address their realm of expertise.

On a similar note, the photographers on this project also have diverse backgrounds. While they all work in documentary photography, largely for magazines, their specialties range from wildlife and nature to portraits and scientific imagery. Unfortunately, there are spatial and temporal limits to what a photographer can capture. The fascinating patterns of ocean circulation or the scale of Arctic ice melt are beyond the reach of traditional photography. So, for many of these subjects we use some of the exceptional imagery now obtained from space. Some photographs were taken from the NASA space shuttle, and some from the orbiting satellites that are more commonly used for weather forecasting or research.

Additionally, we have a number of first-person accounts from people at the front lines of climate research that clearly demonstrate how the field has transcended its origin as a dry academic discipline.

In putting this book together, we hope to impress, educate, intrigue, and maybe motivate. Please let us know if we have succeeded.

Gavin Schmidt
Joshua Wolfe
New York, 2008

CLIMATE CHANGE

INTRODUCTION

Gavin Schmidt and Joshua Wolfe

Whether the weather be cold or whether the weather be hot,
We'll weather the weather whatever the weather,
Whether we like it or not.

—Anonymous

Climate is what you expect, weather is what you get.

—Mark Twain

In writing a book about climate change, it's probably a good idea to start with what we mean by *climate*. A common definition—"weather conditions prevailing over an area or over a long period"—reflects the situation of forty years ago, when climatology solely referred to a very necessary, but rather dull, study of daily temperature and rainfall statistics, and their collation into averages. This definition, however, does little justice to the expansive modern concept of a climate system that incorporates the oceans, atmosphere, biosphere, and polar regions and describes the multiple interactions between these distinct physical entities. While the weather in the atmosphere or the eddies in the ocean are quite variable and even chaotic from day to day, the average conditions in a given location are relatively stable and can be explained and understood in physically consistent ways.

Climate, then, is the average condition of all these environmental components over a period of time. The period needs to be long enough, say thirty years, to smooth out some of the variability associated with the weather day to day and year to year. However, the average condition alone is not enough to define the climate. We also need a description of the variability over the same period—the frequencies of a cold winter or strong rainstorm, or the magnitude of the seasonal cycle—which is also part of a region's climate. The climate can be thought of as all the statistics of the weather (or of the sea ice, or the ocean, or the biosphere), but not the particular sequence of events in any one season or year.

1

If climate is the sum of our expectations, *climate change* is an alteration in those expectations. However, climate change is not limited to alterations in the global mean temperature or rainfall. For example, global warming describes the ongoing rise in mean surface temperatures across the planet, but global climate change encompasses not only global warming but also the occurrence of drought and the shifts in ocean currents or atmospheric winds. Although climate change cannot be seen in any one particular storm, heat wave, or cold snap, it is found within the changing frequency of such events.

We take for granted the ways in which we have adapted to the current climate. After all, the climate as we know it has been relatively stable for hundreds of years. The kinds of crops that are grown, the capacity of storm drains, and the distance that we allow between a building and the shoreline all depend heavily on the expectation that patterns in rainfall, temperature, and sea level will continue as in the past. This expectation has served us well up to now. But how well will it serve us in the future?

Has the Earth's climate changed before? Yes! The planet has seen many different climate changes in the past. Some 700 million years ago, geological evidence indicates that our planet may have been completely glaciated, a condition referred to as "snowball Earth." During the extremely hot Cretaceous period 100 million years ago, dinosaurs ruled a world in which the polar regions were ice-free and heavily forested. More recently, during the last major ice age that ended only about 20,000 years ago, much of Europe and North America was covered with ice that was kilometers thick.

Why did the Earth's climate change in the past, before we humans had any measurable impact on it? Those changes were driven by slow movements of the continents that changed the ocean currents, by asteroid impacts that filled the atmosphere with smoke and dust, and by wobbles in the Earth's orbit that made the summers warmer and the winters colder. These factors (among others) led to dramatic extinctions and spurts of evolution. However, since the time agriculture was invented, animals were domesticated, and our earliest cities were built, all of human civilization has existed in the relatively stable climate of the Holocene epoch, which started around 10,000 years ago. No farmers were displaced when the now-petrified forests of Axel Heiberg Island in the high Arctic first succumbed to the ice some 50 million years ago. No human cities were drowned when sea levels rose 120 meters (400 feet) at the end of the last ice age. What makes climate change different this time is that, over hundreds of years, our modern industrial society has adapted, albeit imperfectly, to the current conditions. Our climate has influenced where we have built our cities, where we plant our crops, how we travel, what we eat, and sometimes, how we die.

Throughout our planet's 4.5 billion years of existence, Earth as a whole has been indifferent to its average temperature. Life on Earth, as it has for several billion years, will eventually adapt to any new situation. But for our particular species, with its huge investment in the status quo, that fact is probably not too comforting. The issue of climate change today is not that the current climate is somehow ideal or perfect, but that it is the one we are used to. Given enough time, we could probably adapt to almost anything. But could we adapt if the climate changed quickly? Would we have enough time?

A medical analogy is illustrative. A doctor can examine our symptoms, try to diagnose our condition, and suggest treatments if the prognosis is not favorable. The success of modern medicine shows clearly that, even when medical knowledge is not perfect, it can still be useful. This is also true for climate scientists studying the Earth—the science is imperfect, but still useful. Drawing from that analogy, we have organized this book into three parts that describe the symptoms of climate change, the diagnosis and prognosis, and suggestions for potential cures and treatments.

The symptoms of climate change can be seen on land, in the oceans, in the stratosphere, at the poles, and near the equator. They can be seen in temperatures, rainfall, winds, plant and animal behavior, and in observations at the local level as well as in images derived from satellites in orbit around the Earth. Diagnosing what the symptoms mean is a job for theorists and modelers who attempt to place these changes in a consistent physical framework. The details are not perfectly described—ambiguities will always exist. But the overall conclusions are robust: much of what is happening is clearly the result of human activity.

Since human activity is not about to cease, or even stop growing, the diagnosis that humans are impacting climate has real consequences. If we take a business-as-usual approach and change very little, those consequences are likely to be serious. Even with significant effort to reduce human impacts, we have no guarantee that dangerous interference with the climate won't occur. Given the nature of the problem and the diverse and complex sources of greenhouse gases, no one simple solution will be possible. But the set of solutions that will be most effective and efficient is still unclear.

In many ways the problem of human-induced climate change is unique: it is global, it will affect the planet for decades to centuries, and it is complex, imperfectly understood, and has the potential for truly dramatic consequences. However, human civilization may have solved one other environmental problem that shares all of these characteristics, albeit on a smaller scale: stratospheric ozone depletion. This story has both connections to, and lessons for, the climate change problem.

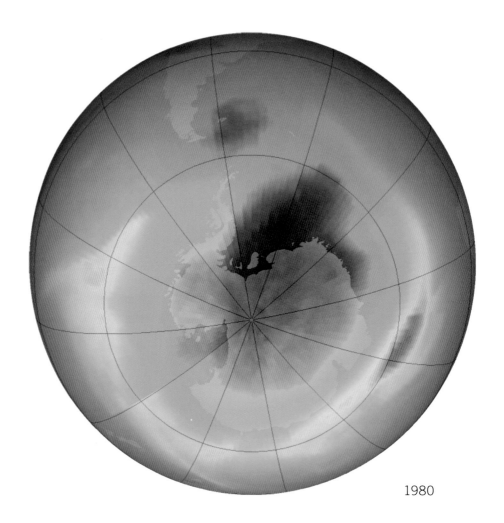

1980

FIXING AN OZONE HOLE

Fifteen to thirty-five kilometers above the Earth's surface, the ozone layer in our stratosphere is made of the "good" ozone that protects life on our planet from the Sun's powerful and dangerous ultraviolet rays. Our atmosphere's shielding layer of ozone exists due to a delicate balance between ozone production from solar ultraviolet rays and ozone loss through the Earth's atmospheric chemistry. In the 1930s, chemists invented a new class of compounds that could be used as a coolant in refrigerators and air conditioners. These compounds, collectively called chlorofluorocarbons (CFCs), consisted of chlorine and fluorine atoms attached to a central carbon atom, forming a large and stable molecule. They were chemically inert, cheap, and apparently harmless, and so had immediate commercial appeal. CFC production increased exponentially over the next four decades. Then in 1973 James Lovelock discovered that almost all of the CFCs ever produced were still sitting in the atmosphere. The CFCs weren't being removed by any natural process. Sci-

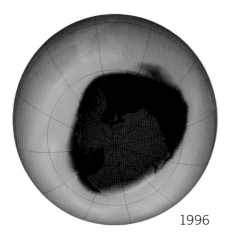
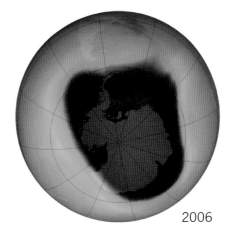

1996   2006

entists soon realized that because of the long life of these compounds, significant amounts of CFCs would make their way up to the stratosphere. There, subjected to the intense ultraviolet radiation, the CFC molecules would break down, releasing chlorine atoms. Chlorine on its own is extremely reactive, and when placed in contact with the ozone layer, as Sherwood Rowland, Mario Molina, and Paul Crutzen suggested, it would catalyze ozone destruction while not being used up itself. This research, which later won the Nobel Prize, was soon confirmed. Concern for the health of the ozone layer grew.

As with the climate change issue, there was uncertainty about what ozone depletion meant. Was it a problem at all? How large would it be? How difficult to fix? The problem spurred enormous amounts of research into the details of stratospheric chemistry to see if the basic mechanism would still be important once a host of other factors were brought in. Models were developed to project what would happen. CFC producers funded disinformation to try to downplay the problem. Some even claimed that CFCs couldn't possibly reach the stratosphere, because CFCs were heavier than air. This claim persisted even after CFCs had already been measured there! (In fact, the atmosphere is so turbulent that all the long-lived gases are well mixed from the surface up to a height of 80 kilometers [50 miles]).

Observations soon confirmed that the models were reasonable and the ozone layer was thinning; however, progress on fixing the problem was slow. In 1985, while international efforts to control CFCs were just getting under way, observations from Antarctica indicated that, instead of the expected gradual decline in ozone, very large and dramatic declines were occurring in the Antarctic region. A "hole" in the ozone layer had in fact developed in the early 1980s, but the measurements were so out of line they had initially been dismissed as a suspected error with the sensor. This phenomenon had not been predicted by any of the models. Scientists soon discovered that the ozone hole was related to small ice particles that form at very cold temperatures in the polar winter vortex. In the spring, these

OPPOSITE AND LEFT: The evolution of the Antarctic ozone hole. Ozone depletion is enhanced above the South Pole because of the existence of polar clouds, which only form in very cold conditions. The ice crystals within these clouds provide extra surface area for the chemical destruction of ozone by the chlorine derived from CFCs. In 1980, a diminuation in polar ozone was barely noticeable. By 1996 it was very noticeable indeed. The ozone hole in 2006 was the largest yet seen, covering some 23 million square kilometers (10.6 million square miles). Now that CFCs have been controlled and are no longer increasing in concentration, ozone recovery is expected to start in the next few years. The amount of natural year-to-year variability will make it difficult to be sure the ozone layer is on the mend for a decade or so.
NASA EARTH OBSERVATORY

ice crystals provide surfaces where more of the damaging chlorine atoms could be released.

After the discovery of the ozone hole, international efforts to reduce CFCs redoubled. By 1987, the Montreal Protocol on Substances That Deplete the Ozone Layer was negotiated. Since then, 191 nations have signed the protocol in an exceptional example of international cooperation. This treaty led to the eventual phasing out of all CFC production, first in the developed world and then in developing countries. Concentrations of CFCs have since started to fall. Although the CFC decrease has not yet produced a reduction in the size of the ozone hole, recovery is expected within the next few decades.

The ozone story is connected to climate change in a number of ways. First, CFCs, in addition to depleting ozone, also are significantly more powerful greenhouse gases than carbon dioxide on a molecule-per-molecule basis (see Chapter 6). Second, changes in stratospheric ozone themselves have climate impacts: globally, ozone depletion has had a small cooling effect on the climate, and the ozone hole specifically is implicated in changes in wind patterns around the Antarctic (Chapter 1). Finally, the reactions that govern polar ozone destruction are affected by colder temperatures, and as the stratosphere cools because of increases in carbon dioxide (this might seem paradoxical, but see Chapter 6 for details), this cooling is likely to push the recovery of the ozone hole back a decade or so. (Occasionally, the two stories have become jumbled in the popular mind, and the question "Isn't global warming caused by the sun coming in through the hole in the ozone layer?" is sometimes asked—the answer is no.)

Apart from these physical connections, the issues of ozone depletion and climate change are connected on political and practical levels. Most importantly, at no time was the science of ozone depletion ever certain. There were (and are) always unanswered questions. Yet the knowledge we did have demonstrated that a problem and a solution existed. Then, as now, extreme exaggerations of the costs of fixing the problem were commonplace. In dealing with the solutions, individual efforts to boycott aerosol sprays using CFCs as a propellant were important in raising awareness. But significant action only occurred within the context of international agreements and with the participation of the major CFC-producing businesses such as DuPont in the United States and ICI in the United Kingdom. Finally, do not forget that the models did not predict the ozone hole. In that case, the models were incomplete and they unknowingly underestimated the scale of the problem. Keep this in mind when reading the prognosis in Chapter 8.

Overall, the lesson learned from the ozone problem is positive. Despite scientific uncertainty, despite reluctance from vested interests, and despite the global equity issues that surround any international agreement, appropriate action was taken, in small steps at first, and then more quickly as the science became clearer and the economic costs decreased with new research into CFC alternatives. It is not beyond imagination that a similar path could be followed in dealing with climate change.

SCIENTIFIC PRACTICE

> **The great tragedy of science—the slaying of a beautiful hypothesis by an ugly fact.**
>
> **—Thomas Huxley**

Science has proved the most reliable and self-correcting method ever devised by humans for finding empirical truths about the real world. Science tests hypotheses rigorously, and mercilessly tosses aside hypotheses that cannot explain the data. The more often a hypothesis is tested and not rejected, the more reliable we judge it to be. Those scientific theories that make interesting predictions gain credence if they are successful.

Over time, a solid foundation of knowledge builds up that has been tested again and again. This body of knowledge, or consensus, may be thought of as the best available synthesis of what is understood on that topic. However, a consensus is not inviolable. Parts of many ideas once thought reliable have indeed been rejected in the light of new discoveries. For instance, Newton's laws of motion were accepted for two hundred years before being shown to break down at very large and very small scales. In the early twentieth century, Newton's laws were replaced by quantum mechanics and general relativity at those extremes (though they are still used very successfully in most everyday applications, such as designing airplane wings or weather forecasting). Sometimes valid ideas, such as Alfred Wegener's theory of continental drift, take time to be accepted. However, once it is clear that a new theory explains observations better than the previous theory and that its predictions have been validated, the scientific community will switch over very quickly. For continental drift, that point came in the 1960s, when unequivocal evidence of sea-floor spreading at mid-ocean ridges gave rise to the theory of plate tectonics, suddenly making sense of many different anomalies, including those highlighted by Wegener decades before. It is this openness to change that distinguishes science from dogma.

Scientists must therefore be professional skeptics, always examining how theoretical assumptions are constructed and how observations are interpreted. Scientific statements that reach consensus level do so because they have withstood many such tests from some of the brightest people around. These statements might still be wrong, but with every test they pass, that becomes less likely.

Many scientific issues, from genetic engineering to climate change, are controversial because they have significant political, economic, or ethical implications. These implications have led to a welcome interest in climate change from the public as well as to an unwelcome abuse and distortion of scientific results for political purposes. Neither attention can be divorced from the other. So, in the public eye at least, climate science appears far more politicized and polarized than it is among climate scientists.

A flip side to this perceived politicization of climate science is the use of scientific language and results by politicians and pundits to "conduct politics by other means." This "scientization" of the political debate has led to lawyers and political pundits discussing the ice age cycles of a quarter million years ago, or fifteenth–century tree rings, as if these past events had some relevance for current environmental policy. While these issues are interesting scientifically (and are covered in this book), their use in modern political contexts is designed to lead to confusion rather than enlightenment.

In any subject as vast and complicated as climate, there will always be anomalies and apparent contradictions in observations, theories, and models. Over time many of these issues work themselves out. Possibly the data were corrupted by a nonclimatic influence. Maybe the interpretation was revised in light of new information. Or the analysis was corrected. Or a new piece of physics was found to be important. At any one time, scientists have many problems within a theory, model, or set of conclusions for which the resolution has not yet been found, and for some, it may never be. A determined critic can often pull together a compelling collection of caveats and contradictions that seemingly turns conventional scientific wisdom on its head. The Internet and the op-ed pages abound with examples of this, on subjects as diverse as evolution, tobacco addiction, and mercury levels in fish.

The problem with these pick-and-choose arguments is that they are constructed to support an already existing view—for instance, that climate change can't possibly be due to human activity. Their creators selectively ignore the far more compelling evidence that contradicts their point of view. This salad bar approach to scientific facts turns normal scientific practice on its head. Like a jury a serving on a trial, scientists are supposed to keep an open mind and come to the conclusion after examining all the evidence, not before.

Partial and distorted readings of scientific findings to support a predetermined political or moral position are nothing new. Regardless of the subject, the techniques used are often the same: distorters overinterpret or misinterpret true but irrelevant facts. They set up straw man arguments that no one has advocated so that they can easily be knocked down, using these as substitutes for similar-sounding ideas of well-supported science. They give non-peer-reviewed opinions as much weight as the assessment of thousands of scientists. They use specific cherry-picked examples to contradict general conclusions—for instance, using the information about one mountain glacier that is advancing to counter the fact that the vast majority are not.

What is the antidote to this unscientific thinking? It is not for scientists to simply dismiss such flimsy claims and argue from authority that the conventional view is correct. On the contrary, scientists need to use the interest and curiosity piqued by these arguments—and more important, the underlying concerns they reveal—as an open door to the public and a chance to demonstrate why scientists have come to the conclusions that they have. On their own, fuller explanations to the public, such as this illustrated book, will not solve the underlying political issues. After all, a general understanding that climate change is happening doesn't translate into everyone agreeing what to do about it, but it does improve the trust that the public has in science. As in so many areas, the answer to bad information is better information.

This book does not attempt to debunk every contrary notion in circulation Instead, we focus on the reasons why scientists think the way they do. Our Further Reading section lists online resources that are reliable and can be consulted for discussions of nonscientific ideas that we do not specifically cover. Where we find true uncertainty, such as in the potential impact of climate change on extreme events, we make that clear. Where the science is not yet complete for all aspects of the specific problem, we describe what is known and why.

What should be clear from reading this book is that we have concluded, as have scientists, assessment panels, and national academies all over the world, that human-induced climate change is ongoing and has the potential to create dangerous consequences for human society if we continue down a business-as-usual path. We do not, however, advocate for any particular solution to this problem. These potential solutions, both technological and political, are discussed in the last section of the book, but decisions on which options to pursue are fundamentally political. While these decisions can be guided by science, they cannot be determined by science alone. We hope that our contributions to this debate—both scientific and photographic—will be a useful guide.

ON COMMONLY USED TERMS

When I use a word, it means just what I choose it to mean—neither more nor less.

—Humpty Dumpty in *Through the Looking-Glass*,
by Lewis Carroll

Scientists often use words in a precise or technical sense that differs from more colloquial use of the same words. To avoid any potential confusion about the words we use in this book, we define some of the most common concepts here.

aerosols: Aerosols refer to almost any particles or droplets in the atmosphere that aren't clouds. Aerosols include dust, pollen, sea salt, smoke, and soot particles. Less noticeable perhaps are the sulfate droplets (sulfuric acid) that come from power station exhausts, which contain sulfur dioxide gas and contribute to acid rain. Other aerosols form part of the haze that we see over polluted cities. Yet other aerosols form the bluish haze over vegetated regions that, for instance, give the Blue Ridge Mountains of Virginia their name. Some aerosols are natural, and some are generated by human activity; all are heavily involved in climate (see Chapter 6 for more details).

albedo (reflectivity): Albedo is the fraction of the Sun's light that is reflected from the planet. Seen from space, the Earth looks bright, not because of the city lights dotted across the surface, but because of the reflected light of the Sun. Most of that reflection is from the bright clouds, but some of it is from the snow and ice in the polar regions, and some from land surfaces such as fields and deserts. For the planet as a whole, the albedo is about 30 percent. For the darker ocean it is about 6 percent, and for fresh snow it is an impressive 98 percent (hence the need for sunglasses on the ski slopes!). Changes in ice, clouds, or the land surface can affect the albedo and thus directly affect the amount of sunlight that comes in. If ice melts because it's warming, this change in the albedo means that more sunlight will be absorbed, leading to more warming. This is the well-known ice-albedo feedback (see the definition of *feedback* below).

anthropogenic: From the Greek words *anthropos* (human) and *genesis* (to create), anthropogenic means that the cause of something is due to human activities. This word does not exclusively refer to greenhouse gas emissions; it can also refer to atmospheric pollution, deforestation, and urban sprawl. It is not to be confused with anthropomorphic, which means to take the shape or form of humans, a term more appropriately applied to the environment of *Animal Farm*, rather than the impact of farm animals on the environment.

business as usual: The scenario in which human society does not make any allowance for climate change in its decisions about emissions, energy efficiency, or technology. It covers a large range of possible and uncertain futures, and is usually invoked as the worst-case scenario.

feedback: The concept of feedback is at the heart of the climate system and is responsible for much of its complexity. In the climate everything is connected to everything else, so when one factor changes, it leads to a long chain of changes in other components, which leads to more changes, and so on. Eventually, these changes end up affecting the factor that instigated the initial change. If this feedback amplifies the initial change, it's described as positive, and if it dampens the change, it is negative. The classic ice-albedo feedback is a positive feedback: snow and ice melt as the planet warms, and because they reflect more sunlight (have a higher albedo) than the now-exposed ocean or land, less solar energy is reflected, further warming the planet. Positive feedbacks in the absence of mitigating factors can lead to runaway effects like the ear-piercing squeal you hear when a microphone is placed too close to a loudspeaker. Luckily, in the Earth's climate, there are enough dampening effects to keep the amplifications from completely running away to the extremes seen on the super-greenhouse planet Venus.

fossil fuel: A small fraction of the carbon in the biosphere gets trapped in shallow seas, soils, and swamps. Over geological time (many millions of years) this organic matter can be transformed into oil, coal, or natural gas through the workings of temperature and pressure. These fossil deposits contain a huge amount of concentrated energy that we've been clever enough to find and exploit. But exploiting these reserves has released the carbon that was stored over millions of years into the climate system over a matter of decades. The resulting perturbations to the carbon cycle are a big focus of this book.

global average: Since the concept of averaging is intrinsic to the definition of climate, it's worth explaining what is meant by a global average and how it is calculated. Climate scientists are generally interested in the overall global picture, not necessarily what is going on in one particular spot. They therefore group the information from each individual data point by taking an average (or mean), which can be weighted to account for the area represented for each point (so that a few regions with lots of points don't dominate the result). How many points do you need to estimate a global average? It depends very much on what is being averaged. For some quantities, such as the concentration of carbon dioxide and methane in the atmosphere, the answer is not very many. Such long-lived gases are well mixed, meaning that their concentration doesn't vary much from place to place; therefore only a couple of observations are needed to get very close to the average value. Temperatures vary on smaller scales and so more points are needed to get an accurate average. There is subtlety in how this is done: global average temperature change isn't the change in the average of all temperatures, but rather is the average of all temperature changes. If that seems puzzling, think about this: when you look at a weather map, you can usually see that if one area is particularly warm or cool, then a nearby area has a similar anomaly, even if the absolute temperatures are very different. This means it's much easier to say whether a region is warmer or cooler than normal than it is to specify its absolute temperature. On a monthly or yearly basis, it turns out you only need a couple of hundred points globally to get close to the true average temperature anomaly. For rainfall, however, the spatial scales are much smaller, such that it can be raining heavily where you live, but a couple of towns over it might not be raining at all. Therefore, you need a really dense network of observations to get a good sense of the average. Luckily, satellites are pretty good at collecting these measurements, even if there aren't enough weather stations. Finally, note that statements about global averages don't mean that every area behaved the same way. The globe can still be warming on the whole even if a few places cooled, as long as the warmer places outnumber the cooler ones.

greenhouse gases: Any gas that, by an accident of chemistry, happens to absorb radiation of a type that the Earth, by an accident of history, would like to lose. All solid bodies emit heat radiation that depends on their temperature. The Sun is very hot, and its heat energy is in the form of visible and ultraviolet light. The Earth is much cooler and gives off energy in the infrared. Atmospheric gases absorb only particular kinds of energy, depending on their structure (the number of atoms in the molecules, the strength of the chemical bonds, their symmetry). Greenhouse gases are those that absorb in the infrared and thus make it more difficult for the

Earth to cool, making the surface warmer than it would otherwise be (see Chapter 6 for more details).

humidity: It's not the heat . . . , but it is closely related. Water is a fundamental component in the climate system—in liquid form in the oceans, as ice in the polar regions, as both ice crystals and liquid droplets in clouds, and as a gas (water vapor) in the atmosphere. The total amount of water vapor in the air is called the *specific humidity*, which varies enormously from the tropics to the poles and throughout the atmosphere. This water vapor is colorless and odorless, but we know when it's there and when it's not, mostly because of the *relative humidity*. For instance, a hot day in summer with a relative humidity of 90 percent is extremely uncomfortable, because it makes it difficult for sweat to effectively cool your body. The percentage is a measure of how close the air is to the *saturation humidity*—the point at which no more water can be added without condensing into liquid water. A related measure is the dewpoint, which is the temperature the air must reach before condensation begins. A key piece of physics is that the saturation humidity increases very quickly as temperature rises. As it gets warmer, the total amount of water vapor in the air can increase by about 7 percent per degree Celsius (4 percent per degree Fahrenheit). Since the processes that remove water from the atmosphere—clouds and rainfall—depend on the condensation of water vapor and the relative humidity, a good approximation in climate science is that relative humidity is quite stable, even though the absolute amount of water vapor in the air can change enormously.

Intergovernmental Panel on Climate Change (IPCC): Formed in 1989 by the United Nations and the World Meteorological Organization, the IPCC assesses the state of climate change science every few years and provides advice to policy makers and its member governments. The first report was issued in 1990, with follow-on reports in 1992, 1995, and 2001. The reports are written by scientists from across the world, and each report goes through multiple levels of open and expert review before the text is finalized. The reports are issued in three sections: Part I deals with the underlying physical science; Part II deals with the impacts of climate change on ecosystems, agriculture, and human health; and Part III deals with options for mitigating (preventing) and adapting to the changes. Because of the large number of scientists involved in writing, editing, and reviewing these documents, the IPCC reports have earned a status as the consensus of the scientific community. The most widely read are the summaries written specifically for policy makers, which are agreed on line by line with member governments so that there is no ambiguity in what the scientists are saying. The Fourth Assessment Report (denoted as AR4) was released in 2007. The complete reports are available at www.ipcc.ch.

models: Climate models run the gamut from simple back-of-the-envelope calculations, to macros that can be put into a spreadsheet, to the very complex General Circulation Models (GCMs) that have hundreds of thousands of lines of code. In other fields, such as statistics or economics, models often refer exclusively to statistical models, in which two or more data series are analyzed and one is explained in terms of the others. Although useful, a fundamental difference exists between that kind of empirical model and the physics-based models that are so often discussed in climate science. A physics-based model is built on the more basic principles of conservation of energy and mass, the laws of fluid dynamics, and the theory of atmospheric radiation. They also include, in the case of GCMs, some empirical approximations of processes that can't be modeled exactly. Because the totality of this physics is more fundamental than an observed correlation, these models are used successfully to explain climate changes in the distant past, as well as to make projections for the future. Models are often an intrinsic part of a theoretical explanation, particularly in complex systems such as the climate. They are best thought of as a quantification of all of our best estimates for how the system works.

radiation: In popular conversation, radiation is a bad thing, evoking memories of 1950s "duck and cover" campaigns and the threat of nuclear annihilation. But in climate science, radiation is a basic mechanism of energy transfer from the Sun to Earth (mostly as visible shortwave radiation), from the Earth to space (infrared longwave radiation), and within the atmosphere. The greenhouse effect itself is fundamentally a block on certain kinds of longwave radiation. So, just as there is good and bad cholesterol, think of radiation in the climate context as "good" radiation.

theories and hypotheses: To a scientist, a theory is much more than a hunch or the random musings of a guy in a bar. Theories of gravity, storm formation, or ocean circulation often are rigorous and extremely detailed expositions of the underlying physics. We don't just have a hunch about why the Gulf Stream is as strong as it is, or why there isn't an equivalent current off the West Coast. We have theories backed by equations, strong matches to observations, and a history of successful predictions. Scientific hypotheses are much closer to the common idea of a guess. Good ones will pan out and may become fully worked-out theories, though more often, they will be discarded when shown to be inadequate at explaining the observations.

units: Most scientists are schooled in the use of the metric system and are comfortable using it. However, many readers do not have an instinctive grasp of what the metric scales mean. This is particularly true for temperature measurements, which in the United States are invariably understood better in Fahrenheit than Celsius. In this book we use the metric system by default, but where necessary, the units are translated into the commonly used units in the United States. For reference, temperature changes in Celsius are multiplied by 1.8 to get the value in Fahrenheit, though in our conversions we usually round off the numbers. For an absolute temperature in Fahrenheit, multiply the value in Celsius by 1.8 and then add 32. One mile is approximately 1.6 kilometers; 1 meter is just over 3 feet; and a metric ton (1,000 kilograms) is just over 2,200 pounds.

PART I

SYMPTOMS

CHAPTER I

TAKING THE TEMPERATURE OF THE PLANET

Peter deMenocal

Mais ou sont les neiges d'antan?
(But where are the snows of yesteryear?)
> — François Villon, "Ballade des Dames du Temps Jadis"

"**31** December, 1768: No one can recall such a mild Autumn: the ground is as green as in the Spring, and today I have picked sufficient young nettles, dandelions, and other herbs to cook green cabbage tomorrow, which is New Year's day."

A colorful mix of meteorology and domestic concerns is typical of weather diaries kept by diligent observers for centuries. This example, from the Stockholm Observatory in Sweden, is not unusual, but it does pose problems for those interested in climate change. For instance, exactly how mild was that autumn and how might it compare to the autumn of 2007? To answer such questions and others like them, these qualitative descriptions are not sufficient—quantitative measures are required.

Galileo Galilei developed the first thermometer in the late 1500s. The "thermoscope," as beautiful as it is imprecise, is an elegant liquid-filled glass cylinder containing several colorful, sealed glass bulbs that rise and sink with changes in temperature as their density relative to the liquid changes. More accurate measurements became available two centuries later, when German physicist Daniel Fahrenheit developed the sealed mercury thermometer in 1714 and the temperature scale that bears his name. As with many scientific advances, this new way of reducing nature to numbers led to a new way of viewing the climate. No longer was the difference between one year and another simply a qualitative change—warmer, cooler, wetter, drier—but a difference that could be reliably quantified. These records gave

A modern Galileo thermometer. As the ambient temperature changes, the density of the water in the thermometer changes (decreasing as it gets warmer). Each glass sphere has a different density that is tuned to match the water's density at a particular temperature (seen on the tag). If a sphere floats to the top, it's density is less than the actual water density, and so the temperature is colder than the value on the tag. The actual temperature can be estimated as being between the temperature of the lowest floating sphere and the highest sphere that is not floating. In this image the temperature is between 66°F and 70°F (19°C to 21°C). © JOSHUA WOLFE

19

A technician checks the automated weather station on Bonaparte Point, near Palmer Station on the Antarctic Peninsula.
© GARY BRAASCH

rise to the statistics of weather and, eventually, to the possibility of detecting subtle changes in climate.

Armed with this new measurement device, weather stations were established in many European cities by the mid-1700s, marking the start of instrumental records of surface air temperature change. Some of the longest instrumental surface temperature records are from central England (1659), Basel, Switzerland (1755), and indeed Stockholm (1756). These stations are very few in number, but since their measurement extend more or less continuously over 250 years, they have become extremely valuable in judging long-term changes in climate. Maintained in the face of war, political upheaval, and economic collapse, "neither rain, nor sleet, nor gloom of night" kept the many generations of observers from their appointed tasks.

Since 1850 or so, a sufficiently large number of stations have covered about 80 percent of the globe, making reliable estimates of the global surface temperature of the Earth possible. Currently, thousands of meteorological stations gather data used to calculate changes in Earth's surface air temperature year after year. However,

it is difficult to put the data from these different stations together to estimate historical global
average temperature change. Each temperature
station has its own "personality," complete with
measurement gaps and other peculiarities such as
the location of the site or the competence of the
observers. These potential problems are often random, and so average out in regional or global scale
averages, although special care must be taken to
account for station moves, instrument changes,
and changes in the way that temperatures are
taken (such as the time of day). Each of these fac-
tors can make a small difference in the mean temperature, which could be misinter-
preted as a climate trend if not accounted for.

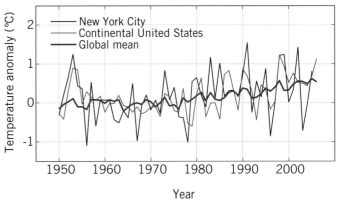

Annual Temperatures at Different Scales

The variations in average
annual temperature vary
more at a single location
(New York City in this
case) than they do over a
continent or over the globe.
In each case, temperatures
are rising, but the signal of
global warming is easiest
to see at the largest scale
(global). DATA FROM NASA/
GODDARD INSTITUTE FOR
SPACE STUDIES

One problem needs particular attention: temperature readings from some urban
settings have had very localized warming trends over time as cool natural vegetation cover is replaced by warmer streets and buildings, and as industry and transport increase the local production of heat. This urban heat island effect can be felt in
many cities and becomes stronger as cities grow. For instance, New York, Phoenix,
and Paris are each a couple of degrees warmer than the surrounding countryside.
The local heat is real enough, but if these central city locations are assumed to represent a wide tract of countryside as well, these data could contaminate the regional
and global mean estimates. The effect can usually be isolated and removed by comparing a given urban temperature record with nearby rural records.

For assessing climate variations, we're most interested in temperature *changes*
over time rather than absolute temperature—that is, the 1°C difference from last
year's temperature is more relevant than whether today's actual temperature is
19°C or 20°C. So each record is standardized by subtracting the average temperature over a fixed base period (commonly 1951–1980, but the choice is arbitrary
and makes no difference to the trends). The resulting temperature *anomaly* record
expresses how temperature has changed relative to that common base period, and
this process is repeated for all stations. The advantage of this approach comes from
the fact that anomalies in different locations are strongly correlated: if London has
a warm summer, then Paris usually has one as well. Closely spaced stations can
be used as a check on station problems (such as an undocumented move) and can
increase confidence in the regional results.

So what do these records show? Most obviously they show that the atmosphere is
a very dynamic place. Combined with the Earth's rotation, variations in atmospheric

Global Warming since 1880

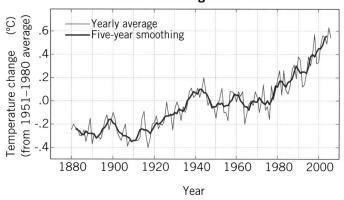

Changes in global mean surface temperature since 1880. DATA FROM NASA/ GODDARD INSTITUTE FOR SPACE STUDIES

pressure cause weather patterns to develop, swirl, and cast off waves of warmer or colder and wetter or drier conditions. These variations can cause large swings from week to week, season to season, and year to year. However, averaged over a continent or globally, these sorts of weather events, as impressive as they are locally, tend to cancel out. For the climate, weather variability is the noise that obscures the underlying climate change. In New York City, the standard deviation from one month to another is around 2°C (4°F); from one year to another it is 0.6°C, whereas for the continental United States it is 0.4°C, and for the globe it is 0.2°C. The larger the area and the longer the period, the smaller the effect of weather.

But over the long term, climate changes can be seen through the noise. At the Stockholm weather station, for instance, temperatures over the last two decades have been about 2°C (4°F) warmer than the 1951–1980 base period, and roughly 1°C (2°F) warmer than the mild "nettles and dandelions" autumn of 1768. Average temperatures in Stockholm in recent decades are the warmest they've been in over 250 years, even after accounting for the urban heat island effect. Note that this increase is much larger than the change from one year to another. For the rest of the world, there is some variation in when the warmest temperatures occurred, but globally, temperatures have risen by about 0.8°C (1.4°F) over the last century to the warmest they've been over the entire record (starting in 1850). Most of this warming has taken place since the 1970s. The warming is stronger in the high northern latitudes than at the equator, higher over land than over ocean, and higher at night than during the day.

This modern instrumental trend is now well accepted by scientists because the record is long enough and the warming is so large that its reality is clear. The warming detected by this global array of instrument stations is also confirmed by many completely independent sources of information.

OCEANS, BOREHOLES, AND SATELLITES

In just the last few years, oceanographers have demonstrated that the surface oceans have warmed by 0.1°C to 0.2°C over the last fifty years or so. This warming is found not only at the surface, but has penetrated to an average water depth of 50 to 100 meters (150 to 300 feet). The measurement of water temperature is exceedingly precise, so this warming signal is well known and significant, and it has been observed in every ocean basin. The oceans cover nearly 70 percent of the Earth's surface and are very important to climate. Not only their spatial coverage is important, but also the fact that water has a very large thermal inertia—it takes the oceans a long time to adjust to changes (see Chapter 3). On the one hand, this inertia implies that measuring temperature changes in the ocean is difficult because the changes are so small, but on the other hand, if global ocean warming is detected, it's a very significant development.

Analyses of ocean temperature are based on millions of measurements made by ships sailing the world's oceans. Coverage in earlier decades was reasonable along well-traveled shipping lines, but quite sparse in the remoter areas of the ocean. Measurements became more widespread when expanding fleets of commercial and passenger ships began routinely to record the intake temperatures of seawater used to cool their engines. Over the years, their voyages have recorded ocean temperature changes for most regions of the world's oceans. An obvious limitation to this data is that some of the more remote areas (such as the South Pacific) have relatively few ship tracks and their temperature history is not definitive.

In the early days, ocean temperatures were measured by dipping a canvas bucket over the side and taking a thermometer reading on deck. Getting a bucketful of water was a feat by itself, as the ship would usually be moving at full speed and some skill was needed to dip the bucket without being pulled overboard. For a laugh, the old hands used to let the young, inexperienced sailors measure bucket temperatures. They would confidently go to the rail with their bucket wondering what the fuss was about. As the light canvas bucket skimmed the ocean surface, it immediately filled and became an impossibly heavy, jerking weight. The poor deckhand would typically be dragged along the ship's rail, his hands blistering on the rope as he struggled to haul the bucket back to the deck. (Speaking from experience, this lesson is needed precisely once.)

In the last few years, a technological innovation has given us the potential to deal with the sampling problem by allowing us to obtain near-global, real-time measurements of world ocean temperatures, not only for the surface but for the subsurface as well. Since 2000, a global array of three thousand drifting floats have been released into the world's oceans to autonomously measure, record, and report (via satellites) ocean temperatures. These Argo floats, named after Jason's oceangoing ship of ancient Greek legend, are about 1 meter (3 feet) long and drift with the ocean currents. Periodically, they sink below the surface by adjusting their buoyancy with an internal bladder, and they record continuous measurements of ocean temperatures to depths of up to 2,000 meters (6,000 feet). They then float back up

to the surface and wirelessly report the results to a data center on land. The record from these devices is not yet long enough to completely specify the trends, but slowly and surely the gaps in coverage are being filled.

Geophysicists also have been trying to test directly whether the Earth's surface rock layer is warming. If the Earth's atmosphere is warming, then the upper ground surface should be warming too, and this heating should have gradually penetrated like a wave into the ground, albeit weakly and slowly, to a depth of many tens of meters.

In a deep borehole, drilled for a water well or any other reason, the temperature in the hole gets warmer and warmer the deeper down you go because the Earth's interior is hot due to the decay of naturally radioactive elements. This geothermal gradient is normally very smooth and can be measured precisely using a special thermometer lowered down the hole. Surface temperature changes due to seasons are large, but since they only last for a few months, the seasonal temperature wave only penetrates to a depth of a few meters below the surface before it is damped. But more persistent temperature changes penetrate to much deeper depths, possibly several tens of meters. The deeper you go, the larger the temperature perturbation needs to have been to be detectable—in some regions, it is still possible to detect cold signals associated with the last ice age 20,000 years ago.

The exceptional warming trend in recent decades can be detected in borehole temperature profiles quite readily. The global warming signal measured in boreholes matches the instrumental record, with warming apparent in almost every region. As with the instrumental data, the greatest borehole warming signals were observed in the high northern latitudes and Arctic regions, where temperatures have apparently risen several degrees above their long-term averages (see Chapter 2 for more details on polar amplification).

These days, measurements are no longer limited to ground (or belowground) observations. Since 1979, weather satellites orbiting the Earth have been measuring a whole spectrum of changes in radiation. An advantage of these satellite-based surface temperature measurements is that they can very nearly measure the entire globe in a few quick orbits. Satellites do not measure temperature directly, but by measuring the radiation at certain wavelengths that emanate from the atmosphere, usually in the microwave band, they can estimate the temperature of different layers of the atmosphere. Microwaves reflect the temperature of vibrating atmospheric oxygen molecules and are measured using Microwave Sounding Unit sensors.

Over the satellite period (that is, since 1979), about a dozen different satellites have been equipped with microwave sounding units, and the records from each need to be tied together to give a climate trend. This can sometimes be tricky since

An Argo float bobbing on the surface of the Pacific Ocean shortly before recovery by the Japanese Coast Guard vessel *Takuyo*.
WWW.ARGO.UCSD.EDU

An artist's rendering of NASA's Aqua satellite, which is currently in low-Earth orbit as part of a constellation of five satellites with multiple sensors that are tracking ocean temperatures, clouds, surface winds, water vapor, and air pollution. The satellites circle the Earth every 99 minutes. NASA/TRW

each satellite's orbit can drift, subtly changing the time of day any one reading is taken, and the instruments themselves can degrade over time. However, despite that uncertainty, the trends from these satellites for the lower part of the atmosphere show very strong correlations and similar magnitudes to the trends seen at the ground.

At slightly different wavelengths, these same satellite sensors have been used to show that the stratosphere (at an altitude of around 20 kilometers [12 miles] and higher) has been *cooling* significantly over the same time that the lower atmosphere has been warming. While this might at first seem contradictory, these opposing temperature trends between the upper and lower reaches of the atmosphere are an important fingerprint that identifies the causes of these changes (see Chapter 6 for details).

Curiously, this upper atmospheric cooling means that the sky is, literally, falling. The upper layers are becoming more dense and are therefore contracting, and so the stratosphere and the layers above it are slowly falling closer to earth. Chicken Little might not have approved, but a positive benefit of this change is that it reduces the frictional drag on the satellites, and thus is helpfully prolonging the lifetimes of the very satellites that observe the phenomenon.

With these temperature changes have come dynamic changes in wind patterns. In both the northern and southern midlatitudes, the westerly winds (which flow from west to east) in winter have been stronger in recent decades. This is changing the storm tracks, particularly in Europe, where winter storms now track farther north than during the 1950s or 1960s, leading to significant drying around the Mediterranean from Spain to the Middle East, and to even more rain in Scotland and coastal Norway. As if Bergen, where it already rains 250 days a year, really needs it . . .

LOOKING FURTHER BACK

One of the more challenging questions surrounding the recent record-setting surface temperatures is how unusual the current trends are in the broader long-term context. The challenge is that our observation period is really quite short relative to the long timescales of global climate change. As we have observed, global thermometer readings only extend back 150 years, and ocean temperature and satellite-based records only reach back a paltry 50 or 30 years, respectively. As compelling and consistent as these temperature records are, they are still too short to assess the magnitude of natural variations in temperature over the last several centuries or millennia.

Luckily there are reliable ways to do this using *proxy* records of past climate change. Proxy records are surrogate indicators of past climate change that derive from natural recorders of climate variability, such as tree rings, corals, fossil pollen preserved in lake sediments, ocean sediments, clam shells, and glacier movements. They aren't direct measures of temperature or rainfall, but they are so closely tied to them that changes in the proxy can give a strong clue to changes in climate.

An example of a proxy record is the use of tree rings to reconstruct past changes in temperature or rainfall. Temperature controls the growing season of some tree species in certain locations very closely. These trees lay down visible rings each year, with light, low-density layers reflecting growth in the warm season, and dense, dark layers marking the cessation of growth in the cool season. Since some trees live for hundreds of years and forests of trees represent a broad spectrum of ages, it's possible to take small borings from dozens of trees in a particular location and to build a single, composite tree ring record that documents changes in tree growth spanning many hundreds of years.

Tree rings are not the only biological proxy that has annual banding. Corals in the tropical oceans produce a hard calcium carbonate skeleton that is also laid down in annual increments. The carbonate is made of constituents that the corals

Scientists Ed Cook and Rosanne D'Arrigo from the Tree Ring Lab at the Lamont-Doherty Earth Observatory at Columbia University. © PETER ESSICK

extract and precipitate from seawater, and these can be analyzed to determine the water temperature in which the coral grew. Some corals grow to be several meters in diameter, and their skeletons can represent a virtual library of many hundreds of years for continuous coral growth in a single spot in the ocean.

A guiding principle in paleoclimate research—the science of reconstructing past climates using proxy measurements—is that proxies such as tree rings must be calibrated against observed variations of climate, and the resulting product must be validated against known historical changes. Only after climate proxies have been both calibrated and validated can the proxies be trusted to provide useful data for analyzing climates of the more distant, preinstrumental past. Tree ring records have been particularly valuable for reconstructing past temperature changes. Statistical methods are used to combine tree ring temperature estimates from different regions of the world to develop global temperature reconstructions that extend up to two thousand years into the past.

Accuracy degrades the further back one goes, but these studies have shown that modern temperatures are the warmest they've ever been in the last four hundred years, and it's likely (though not absolutely certain) that we are in the warmest period globally in over a millennium. The increased uncertainty prior to the year 1600 is mainly a consequence of the small number of available records that are long enough; more and better data are definitely needed. Still, the temperature reconstructions suggest contemporary global warming is exceptional even when compared to longer timescales of Earth history.

Europe appears to have been quite warm during the Middle Ages, but warm periods elsewhere don't seem to have happened at the same time, making the Medieval Warm Period a regional phenomenon. For contrast, coral records provide a very long baseline of past tropical Pacific variability that reaches back many centuries and shows medieval times to have been characterized by cooler conditions. They also show El Niño events becoming both more frequent and more intense after the mid-1970s, including a dramatic and unprecedented freshening of the western tropical Pacific Ocean (see Chapter 3 for more details). Whether this is connected to the long-term temperature rise or is just the noise in a chaotic system is as yet uncertain.

The most interesting result from these reconstructions is not the tracking of the warmest year this millennium but the structure of the temperature and rainfall patterns in space and time. These can often be tied to documented changes in human societies and landscapes that are found in art, oral histories, or archeologi-

Mesa Verde in Colorado is one of a number of Ancestral Pueblo sites in the Four Corners region of the American Southwest. The Square Tower House ruins were abandoned roughly eight hundred years ago at a time of severe drought. © GARY BRAASCH

cal sites. For instance, cold periods during the fourteenth and seventeenth centuries and even as late as the nineteenth century, sometimes called the Little Ice Age, correspond to occasional Frost Fairs on the frozen Thames River where the citizens of London would drink, skate, play sports, and socialize. Some of the very long instrumental records also capture these cold periods, such as the "Year without a Summer" in 1816 in the wake of the Tambora volcanic eruption. Going further back in time, drought reconstructions from tree ring records indicate that severe droughts were coincident with the collapse of the classical Mayan cultures in the Yucatán, Mexico, and the disappearance of the Ancestral Pueblo (Anasazi) culture in the American Southwest (see Chapter 7 for more details).

INTEGRATING GLACIERS

Mountain glaciers are very useful indicators of climate change. Found on every continent and at almost every latitude, these rivers of ice slowly integrate over many years of weather and react mainly to long-term changes. The presence of a glacier in a high mountain valley indicates that the long-term average rate of snowfall exceeds the rate of snowmelt at that altitude. Moreover, air temperatures there must be on average below the freezing point of water. As anyone who has climbed even a modest mountain will have observed, it's colder at the top. The atmosphere is compressible, and the higher you go, the less air (and pressure) there is above you, so the air expands, cooling at the same time. This is the same effect (but in reverse) that occurs as you pump air into a bicycle tire—as the air compresses, it warms up considerably. The atmosphere cools at a rate of about 7°C per kilometer of altitude, so if it's a comfortable 21°C (70°F) at sea level, one might expect to find alpine glaciers at about 3 kilometers of elevation.

More precisely, the existence of the glacier is determined by the delicate balance between snowfall accumulation on top in winter and melting that occurs at the lower elevation zone in front of the glacier, usually in summer. The melting zone is commonly marked by a terminal moraine, a massive, curving rock pile that marks the position where the flowing glacier disintegrates and drops the load of pebbles and rock that it has scraped from the valley. The Little Ice Age cooling events are marked by terminal moraines at levels roughly 300 meters (1,000 feet) below their present level in alpine glacier valleys across Europe and Scandinavia. This difference tells us roughly how much cooler it was during the Little Ice Age. Using the average atmospheric cooling rate with height mentioned above, the Little Ice Age in those regions works out to have been about 2°C (4°F) cooler than today. This is similar to the local cooling inferred from the proxy records, such as tree rings.

As one of the most visually compelling indicators of climate change, mountain glaciers have been melting and retreating at an accelerating rate in recent decades. This exceptional meltback is observed nearly everywhere around the globe where there are mountain glacier systems, from New Zealand to Patagonia, from Montana to Switzerland. Glaciologists are alarmed that the vast majority of well-documented glaciers are retreating at an accelerating pace, although there are a few exceptions to this general rule—usually in cases where increased winter snowfall has compensated for the increased summer melt, as in Norway. Like the recent rise in surface air temperatures, the magnitudes and rates of recent glacier melting also appear to be unprecedented, at least over the last four hundred years.

Glaciers in the tropics have fared even worse. The tropics are warm and seasonality is low, but glaciers can form at high elevations on isolated plateaus and mountain peaks in the Andes, East Africa, the Himalayas, and Papua New Guinea. At this elevation, temperature changes across the tropics are strongly coordinated, much more so than at the surface, and so coherent tropical glacier retreat is a likely sign of a long-term, tropics-wide warming. Despite their precarious existence, the largest tropical glaciers are ancient, containing ice that is many thousands to tens of thousands of years old. These majestic tropical glaciers silently remind us that

The Athabasca Glacier drains the Columbia ice field in the Canadian Rockies and has retreated more than a kilometer since the early twentieth century. The retreat has accelerated since 1980.
© JOSHUA WOLFE

This image perfectly captures the dynamism of glaciers literally pouring off the Kukri Hills into the McMurdo Dry Valleys of Antarctica. For a sense of scale, note that the drop from the ice sheet above is over a thousand meters (3,000 feet). © HEIDI GODFREY

OPPOSITE: A stream of glacial runoff from the Quelccaya ice cap in Peru. © PETER ESSICK

past climate swings over prior millennia were never large enough to have melted the glaciers away.

At present, however, tropical glaciers—large and small and on all continents—are disappearing, and they are melting away at an accelerating rate. The melting of the larger, older tropical glaciers is particularly troubling because they provide such obvious evidence that the modern, industrial-age warming is exceptional. Some glaciers appear to be rotting from the inside as meltwater trickles down through ice crevasses, whereas others seem to be literally evaporating away. For example, the small ice cap atop Mount Kilimanjaro in Tanzania has lost 80 percent of its mass, and Hemingway's "snows of Kilimanjaro" are expected to become just a memory in about a decade or so.

The largest tropical glaciers are found in Peru's Cordillera Blanca, or White Mountain Range. The Quelccaya Ice Cap is the largest tropical glacier, covering roughly 44 square kilometers (16 square miles). One of the largest glaciers flowing

from this ice cap has been intensively studied, and the best estimate now is that the glacier will be gone completely by about 2012. The accelerating rate of tropical ice melting is a real concern for Peru, not only because the White Mountains will soon lose their moniker, but because the glaciers store water seasonally for agriculture and hydropower for tens of millions of people. Changes in that storage will exacerbate summer water stresses in the lowlands.

UNEQUIVOCAL WARMING

The collection of warming indicators discussed in this chapter demonstrates why the 2007 Intergovernmental Panel on Climate Change assessment described recent warming as "unequivocal." Nineteen of the warmest years on record have occurred within the last twenty-five years. The warmest years globally have been 1998 and 2005, with 2002, 2007, and 2003 close behind. The warmest decade has been the last ten years. The warming is substantially more widespread than in any previous warm decade (such as the 1930s), and the rise appears to be exceptional on even longer timescales. The odds of this sort of clustering, if it occurred only by chance, would be less than one in a billion, which is almost like being the newcomer to a game of roulette and having your ball consistently land in the number 1 pocket, spin after spin. It's exceedingly unlikely that the current warming trend is, well, not a trend.

How does modern warming fit into a broader, Earth history perspective? The last time the Earth was substantially warmer than this for a sustained period was about 3 million years ago, during the early Pliocene, when global temperatures were about 3°C (5.5°F) warmer and geological evidence shows that the Arctic was ice-free, the Greenland Ice Sheet was nearly absent, and sea levels were possibly 20 meters (over 60 feet) higher. Other more recent periods appear to have been slightly warmer, specifically the last interglacial period around 125,000 years ago, when Greenland temperatures might have been 3°C to 5°C (5°F to 9°F) warmer and sea level 4 to 6 meters (20 feet) higher.

The bottom line is that although current warming is not unprecedented in all of Earth history, previous eras that were clearly warmer than today were accompanied by changes (particularly in sea level) that dwarf the variations any modern humans have seen.

Gary Braasch

CLIMATE CHANGE IN THE UNITED STATES

With so much dramatic imagery coming from the polar regions or the Sahel, one might conclude that changes are only happening in remote places. However, the effects of climate change are also being seen in the United States, and Gary Braasch has been recording them for years. Although a photograph cannot show statistics, these seven evocative images capture a variety of impacts ranging from the obvious (heat waves and flooding) to the subtle (the complex reactions of ecosystems), typifying the trends so far.

The first image is an aerial view of Chicago during the severe 1995 heat wave that killed more than 600 people. Longer and more intense heat waves are a clear prediction of global warming, and they have been increasing in frequency. Heat and the associated increases in air pollution, seen clearly as the haze in the photograph, are two of the most significant causes of increased mortality due to climate change. Heat-related deaths spiked during the 1995 Chicago heat wave, and many other people were affected by the cardiovascular impacts of poor air quality.

Dramatic as these heat events seem, the more far-reaching impacts Gary has documented are those associated with changes in rainfall patterns. In the American Southwest, the bathtub ring around half-empty Lake Powell on the Utah-Arizona border is a testament to the ongoing multiyear drought in the region. Ironically, rainfall is increasing farther north, but more frequently it is arriving in higher-intensity bursts that can lead to flooding. Floods such as the one that occurred in Gurnee, Illinois, in 2004 can be caused by intense and prolonged rainfall, but can be exacerbated by river and floodplain management decisions. In Oregon, intense rainfall is often associated with strong El Niño–related storms that can sweep away coastal roads. These storms do not appear with every El Niño event, but the odds of destructive storms increase sharply during those years.

On the East Coast, coastal erosion is driven by rising sea level and the associated increase in storm surges, even if there is no increase in storm intensity itself. As Gary shows in his photograph from Cape Hatteras, North Carolina, erosion has been exacerbated by inappropriate coastal development can that affect beach dynamics.

Some impacts are more subtle, though. Changes in phenology, the timing of natural events, can illuminate the trends in U.S. ecosystems. For instance, the Virginia bluebell is flowering seventeen days earlier than bluebells would have a few decades ago. Similarly, the whitebark pine near the tree line of Mount Washburn in Yellowstone National Park in Wyoming is having trouble coping with higher temperatures. While the tree line is moving slowly up the mountain, a more insidious problem is presented by the expanding range of pine bark beetles. A massive rise in beetle populations has resulted in the loss of hundreds of thousands of whitebarks in some areas, complicating efforts to deal with the threat of non-native blister-rust fungal disease, which can more easily attack water-stressed and beetle-infested trees.

Chicago heat
wave, 1995

Lake Powell,
September 2004

Flood in Gurnee,
Illinois, 2004

An Oregon
coastal road
destroyed by a
strong El Niño–
related storm

Abandoned house,
Cape Hatteras,
North Carolina

Virginia bluebell

A whitebark pine
near the tree
line of Mount
Washburn in
Yellowstone
National Park,
Wyoming

CHAPTER 2

CHANGES IN THE NORTH

Stephanie Pfirman

The ice was here, the ice was there,
The ice was all around.
It cracked and growled, and roared and howled,
like noises in a swound!
                    —Samuel Coleridge, "The Rime of the Ancient Mariner"

As our hunting culture is based on the cold, being frozen with lots
of snow and ice, we thrive on it. We are in essence fighting for our
right to be cold.
                    —Sheila Watt-Cloutier, former chair of the Inuit Circumpolar Council

Looking down at the North Pole on a globe, Arctic sea ice and the Green-land Ice Cap appear as a bright, white blanket surrounded by the dark green tundra of Siberia and North America and the dark blue of the Atlantic and Pacific oceans. This contrast between light and dark is one of the defining features of the Arctic, giving it a special place in the Earth's climate system. As the Earth warms, the snow and ice that make the Arctic bright will melt, exposing the blue ocean waters beneath the sea ice and the land beneath the snow. Even the land will shift toward darker shades of green as the low-lying tundra vegetation that is easily covered by minor snowfalls is replaced with shrubs and dark green spruce that can poke through the thickest snow.

This is important, because darker surfaces absorb more sunlight—in the same way a white hat keeps you cooler than a black one—and the darkening of the Earth's surface leads to more warming, which leads to additional melting. In summer, the almost constant presence of both ice and open water in the Arctic keeps the temperature close to the freezing point. Any additional energy goes toward melting

Gazing out to the Arctic Ocean over a small patch of recently formed pancake sea ice off the coast of Alaska. © PETER ESSICK

45

Multiyear sea ice off the coast of Kaktovik, Alaska. This ice is a couple of meters thick (most is below the surface) and is typical of ice in the late summer, with plentiful freshwater melt ponds on the surface. Floes this thick are becoming more and more rare. © PETER ESSICK

the ice rather than warming up the air; but once the ice has melted, solar energy during the twenty-four-hour summer day can rapidly warm the ocean water and land areas.

The Arctic, north of the Arctic Circle at 66° North, represents only 4 percent of the Earth's surface, but it carries with it the significance of acting as the Earth's air conditioner, keeping the planet cool. The Earth's atmosphere and oceans constantly move excess heat from the tropics toward the North and South poles. Over the last thirty years, however, temperatures have increased almost twice as fast in the Arctic than in the rest of the world. Arctic ice and snow are melting so quickly, and polar ecosystems are changing so fast, that Inuit elders have said "the weather today is harder to know."

Changes are occurring most rapidly along the periphery of the Arctic, eating away at the ice and snow along the edges, while the center is still kept white and cold by the presence of ice. On land, there have been changes in snow: it falls later in the fall and melts earlier in the spring. The permafrost (permanently frozen ground) is warming and thawing, undermining the structures built on it. Glaciers are melting back along the edges, or slipping forward into the sea and breaking up into icebergs, more rapidly

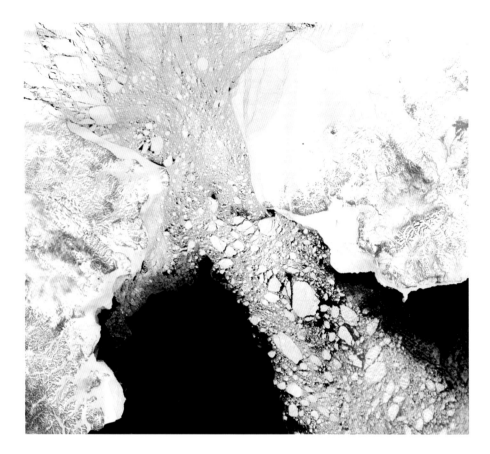

A satellite image looking down over the Bering Strait, which separates snow-covered Cape Dezhnev, Russia, and Cape Prince Wales, Alaska. Open water "leads," seen between the individual ice floes, allow the winds to push against the ice, forcing the Arctic Ocean spring pack ice through the narrow 60-mile-wide strait. NASA/GODDARD SPACE FLIGHT CENTER

than in prior centuries. In the ocean, the sea ice forms later in the fall and melts earlier in the spring, so that coastal seas now have extended periods of open water.

While some of these changes might be helpful—less energy is needed for home heating, open coastal seas are less dangerous for shipping—ice is so much a part of the Arctic that its disappearance destroys the very fabric of the place. Transportation over land is much easier in winter when the land is frozen. Although snow might inhibit the growth of plants and animals in other places, in the Arctic it acts like an igloo to insulate the ground surface, protecting small burrowing mammals from frigid atmospheric temperatures in winter.

Such contradictions and complexities make the Arctic a fascinating place, and the Arctic's connections with the rest of the world make it a critically important one. One crucial difference between the Arctic and Antarctic is that the Earth's northern pole is thin, patchy ice cap drifting on top of the large, current-filled Arctic Ocean. The Earth's southern pole sits on a thick ice cap that in turn completely covers the stable Antarctic continent. Year-round sea ice covering the central Arctic Ocean not only affects how much heat is absorbed in the Arctic, but also cloud cover and the humidity of the atmosphere. The Arctic cools the atmosphere and the deep ocean, adding oxygen to it and driving the circulation and ecology of the deep sea.

Furthermore, the melting of Arctic ice can create another amplifying cycle, since carbon-rich plant matter stockpiled in the permafrost is potentially poised to release the greenhouse gases carbon dioxide and methane to the atmosphere. If this carbon is unlocked through thawing and decomposition it would further speed warming. The stability of global sea level is also dependent on ice stored in Greenland and other Arctic and Antarctic glaciers (see Chapter 3).

SEA ICE

Sea ice forms along the Siberian coast in what is called the "ice factory." As sunlight wanes in the fall and winter, cold winds blowing north cool the ocean waters, freezing a thin skin on the sea surface. The winds and the waves break up the ice into floes, which are patches of ice sometimes kilometers in diameter. As the winds blow the ice for several years across the Arctic Ocean toward Canada, Greenland, and Norway, it continues to thicken by freezing on the underside of the floe each winter. The floes also jostle into each other, often buckling under the pressure and riding up and over each other to form ridges, some many meters thick. As a result, while the new ice along the Siberian coast is thin and responds quickly to warming of the atmosphere and ocean, the older ice along the coasts of northern Greenland and the Canadian Archipelago is thicker and more resistant to change.

Each summer, the sea ice along the shallow coasts and marginal seas melts, leaving only the thick multiyear ice that survives seasonal melting and covers the deep basin in the center of the Arctic Ocean. In the past, this annual variation in the extent of sea ice ranged from 15 million square kilometers at a winter maximum in March to 7 million square kilometers at the end of the summer melt season in September. But satellite images and other observations since 1979 show the regional extent of annual Arctic sea ice has been decreasing at a rate of more than 3 percent per decade. The reduction is greatest in the area covered by at least some sea ice (the extent) during the summer. The summer of 2005 was a record low, but the summer of 2007 set a clear new record for minimum ice extent of only 4 million square kilometers.

The extensive expanse of multiyear ice in the Arctic is unique. The only other place on Earth with perennial sea ice cover is in Antarctica, and there it is reduced from a maximum of 18 million square kilometers in winter to a patchy distribution of only 3 million square kilometers during the Southern Hemisphere summer. With the recent changes in the Arctic, it is approaching conditions more like those found in the Antarctic.

Inuit hunters returning home through light ice near Barrow, Alaska. They hunt for ringed seals, who make their dens in the thickest ice, either in heavily ridged sea ice, or under snowdrifts in land-fast ice stuck to the coast.
© GARY BRAASCH

The decrease in ice extent is troubling, but no less problematic is the fact that the ice is thinning. It is harder to observe ice thickness changes than it is ice extent, but measurements from submarines and other sensors indicate that over the past thirty to forty years the average thickness of sea ice in the central Arctic decreased by about 1 meter to now less than 2 meters (6 feet) thick, about the ceiling height in a typical room. Thinner ice takes less energy to melt, so this means that the remaining ice is becoming more vulnerable. Because the ocean is warmer than the atmosphere in winter, thinner ice also allows the ocean to become a greater source of heat for the atmosphere during the cold season.

The extensive winter freezing and summer melting of the ocean surface plays an important role in the ocean's heat- and salt-induced (thermohaline) circulation (see Chapter 3). When the upper layers of the ocean become denser than the waters below, they sink and flow out of the Arctic into the great oceanic conveyor circulation. Warmer waters from the south flow into the gap left by the sinking, convecting northern surface waters. The Gulf Stream's northern tail, meandering from the Atlantic Ocean northward along the coast of Norway and on into the Arctic Ocean, has warmed in places by more than 1°C (2°F) since the 1990s in comparison with previous decades.

On Greenland's eastern coast, multiyear ice transported out of the Arctic used to be a commonplace feature, flowing rapidly southward along the shore, isolat-

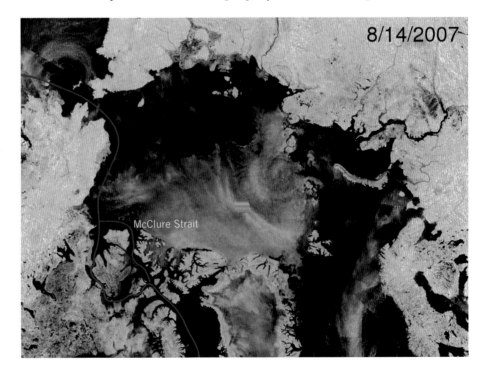

The Northwest Passage was sought for centuries as a shortcut to Asia from Europe by explorers like Sir John Franklin, who found only ice-blocked passages. The first successful transit, by Roald Amundsen in 1904, was via the southernmost route. The journey took over two years and was not repeated until 1944. In August 2007, and again in 2008, for the space of several weeks, this passage was completely free of ice. More surprising still, the McClure Strait also was completely open, something not seen since routine monitoring began in the 1970s. © WILLIAM CHAPMAN

8/14/2007

McClure Strait

ing the coast from the rest of world. From 2001 through 2004 much of this region was completely free of ice in the summer, allowing cruise ships to exploit access to places they had not been able to visit regularly before. As other Arctic passages that had been choked with ice open up, it will allow transportation on a scale never before seen in the region. In past decades, the Northern Sea Route along the coast of Siberia has been open around one month each year, though with substantial differences from year to year. Projections of sea ice concentration imply that this period could triple by the end of the century. While this route is remote for many countries compared with using the Suez or Panama canals, it cuts 40 percent off the shipping time between important markets in northern Europe and the northern Pacific.

On the other hand, the decreasing thickness and extent of sea ice has made it more mobile, and more dangerous in some regions. Because ice is pushed by winds from Siberia across the Arctic Ocean toward the Canadian Archipelago, the flow of thinner and patchier ice may increase through the Northwest Passage. In the past, ice trying to get through the narrow channels piled up quickly, forming ice jams

As the dominant predator in the Arctic and surrounding regions, polar bears are acutely sensitive to large-scale changes in sea ice extent, because they depend on the ice for their survival. © RICK SAMMON

that blocked the flow. Warming could result in unstable ice conditions that are actually less predictable and therefore more problematic.

Although the reduced Arctic sea ice is debatably a benefit to those who just want to pass through, for the marine mammals that depend on the ice, its loss is a significant problem. Diminishing ice means diminishing habitat for polar bears and ice-dependent seals. Polar bears excavate snow caves to give birth to their cubs; these caves are often on land, but sometimes are dug on sea ice. The polar bear's main prey is the ringed seal, and bears use sea ice as a platform to reach and locate ringed seal breathing holes. Locating food in spring is an urgent quest for mother polar bears, which have not eaten for half a year by the time the cubs are born. Ringed seals also breed on the sea ice in snow lairs. When the snow is not thick enough to support a seal lair through the spring, the seal pups can be exposed to the cold and die. Unstable sea ice that breaks up prematurely also forces young ringed seals into the water too early to be reared properly, and they do not survive.

These changes in sea ice will affect not just the seals and polar bears, but will have consequences for the rest of the ecosystem that depends on them. Of all the marine "country food" in the Arctic, the ringed seal is the most critical for the Inuit, for the same reason that the seal is important to polar bears: it is the most abundant seal species in the Arctic, numbering in the millions. Ringed seals are able to create breathing holes in the densest sea ice using the heavy claws on their foreflippers. Since every seal must breathe, these holes are the best place for a bear or an Inuit to hunt for one, even during difficult ice conditions when little other food is available.

ARCTIC LANDS

At the same time that ice is melting at sea, it is also melting on land. We see the thawing and disintegration of glaciers and the degradation of permafrost. Glaciers form by the accumulation of snow, layer after layer, year after year. The pressure of the overlying snow compacts the lower layers, turning them into ice. As the ice compresses and recrystallizes, air spaces are reduced, making the ice clearer. Because of the scattering of blue light waves and absorption of the red, glacier ice acquires a characteristic blue tinge. As snow accumulates, the ice mass becomes sufficiently thick to deform under its own weight and flow downhill. This movement transfers the glacier ice to lower elevations, where it melts or evaporates, or where it meets the ocean and breaks off (calves) into icebergs. A glacier in balance is one in which snow accumulation at higher elevations matches ice loss in lower elevations due to melting and calving. Glaciers retreat when this balance is upset—when there is too little accumulation of snow, or when it melts too much or calves too quickly.

OPPOSITE: Signs along the route to Exit Glacier in Alaska document a 2-kilometer-long recession from its maximum extent in 1815 to today. © JOSHUA WOLFE

A glacier is referred to as "cold" when its internal ice temperatures are well below freezing and the ice is frozen to the ground. In these cases, the glaciers move downhill by deformation from the weight of the ice alone. A glacier is considered "warm" when its ice temperature is at the melting point and water can exist throughout the ice mass, in particular at the bottom. Typically, water at the bottom comes from melting on the glacier surface. The water penetrates into the ice via cracks or vertical pipes called moulins. Surface streams on glaciers are sometimes observed to drain into moulins, allowing the water to cascade down to the glacier bed.

Because water beneath a glacier makes the surface slippery, warm glaciers move by sliding along the surface as well as by deformation. In some extreme cases, water accumulated at the bottom of a glacier greatly reduces friction and the glacier surges forward, thinning and drawing down the reservoir of ice in the accumulation area.

In addition to surging, glacier flow can be destabilized if a blockage is removed— for example, when glacier flow that has been hindered by pressure from adjacent glaciers is released, or when the side drag of the glacier is relieved. When a blocking glacier breaks up or when the glacier front is no longer pinned by islands or fjord margins, a chain reaction destabilizes the glacier that was being held back.

The Greenland Ice Sheet, averaging about 1.6 kilometers (1 mile) thick, comprises the vast majority of Arctic glacier ice volume, and Greenland is experiencing all three types of glacier changes: melting at the surface, surging forward, and breakup by destabilization. Estimates of the total mass of the ice sheet indicate that annual ice loss doubled in the last decade. Since 1979 (when satellite observations first became available), the area of the surface experiencing melting in summer has increased on average by about 1 percent a year in concert with a 2.4°C (4°F) atmospheric warming in the region over the same period. Maximum surface melt extents were observed in 2002 and 2005, when nearly half of the surface of the ice sheet experienced some melting.

But surface melting accounts for only about one-third of Greenland's ice loss; the other two-thirds comes from changes in glacier flow. The flow speed of many of the glaciers draining the southern part of the Greenland Ice Sheet is increasing, resulting in thinning of the edges of the ice cap. These glaciers are speeding up in part because they are able to slide faster as more water is supplied to their bases by the extensive surface melting.

Several Greenland glaciers that terminate in the sea have exhibited dramatic changes. The flow speed of the Kangerdlugssuaq Glacier in east Greenland accelerated more than 200 percent from 2000 to 2005, reaching velocities of up to 14 kilometers (8 miles) per year at the calving front. Along the west coast, Jakobshavn Isbræ, the largest outlet glacier in Greenland, which drains 7 percent of the ice sheet

The Mendenhall Glacier in Juneau, Alaska, photographed in 1894 and again in 2004. As with almost all Alaskan glaciers, and indeed most glaciers globally, the front of the glacier has retreated dramatically because of warmer temperatures during summer and reduced snowfall in winter. RIGHT: NATIONAL SNOW AND ICE DATA CENTER COLLECTION, NATIONAL OCEANIC AND ATMOSPHERIC ADMINISTRATION; OPPOSITE: © GARY BRAASCH

area, experienced an almost 100 percent increase in speed from 1996 to 2005. In the case of Jakobshavn Isbræ, its acceleration was largely caused by the breakup of its ice tongue that extended into the sea. Glaciers with a floating marine section are especially vulnerable to atmospheric and oceanic warming because the ice melts from both the top and the bottom and is easily broken up into icebergs that float away.

Other smaller glaciers and ice fields in Canada, Alaska, and Russia also are melting and retreating. Although they have been declining since the 1960s, an accelerating trend has been seen since 1990. Alaska's glaciers have experienced some of the most dramatic changes: an aerial survey of two thousand glaciers indicated that 98 percent of low-altitude glaciers are thinning or in retreat; just a dozen out of seven hundred named glaciers are advancing. In the past twenty-five years, the ice front of the Columbia Glacier retreated 15 kilometers (10 miles).

Glacier melting is important not only to the Arctic but also to the rest of the world. When water is transferred from land to sea, it raises sea level everywhere. Taken together, if all the glacier ice in the Arctic were to melt, sea level would increase by about 8 meters (25 feet). Melting of sea ice does not result in significant sea level rise because it is already floating and so displaces its own mass, just as when ice floating in a drink melts the glass does not overflow.

A few glaciers, mostly along the western coast of Norway, increased in size from the 1970s until the end of the 1990s, but are now retreating. They advanced at first because they received more snowfall in winter, associated with the increasing trend in westerly winds (mentioned in Chapter 1), than could be melted in summer. In fact, Arctic precipitation is projected to increase, perhaps by 20 percent by the end of this century, because of the higher humidity levels warmer air can support (as discussed in On Commonly Used Terms). But over most of the Arctic, warmer conditions mean that rain, not snow, will become more prevalent. The amount of precipitation and the timing of rain and snow events are of critical consequence. For example, when the ice on streams and lakes has sufficient time to freeze before a major snowfall, the ice is thicker, creating a safer surface for transit by people and animals. Snow insulates the ice from the atmosphere, causing ice to grow more slowly. Recent changes in the timing of snowfall have led to dangerous conditions, with an increase in accidents as people, and animals, break through weak ice.

Another major change that already has been observed is earlier melting of snow in the spring. In Barrow, Alaska, melt onset is about 10 days earlier than it was fifty years ago. As the snow melts earlier, so does the underlying ice, reducing the length of time that lakes and streams are frozen. One of the most notable events in a warmer world will be the loss of year-round ice cover over thousands of ice-bound

The Greenland Ice Sheet
accumulates snow on the
surface and loses mass
through a number of fast-
flowing outlet glaciers
like this one that deliver
icebergs to the sea.
The changing dynamics
of these glaciers are
imperfectly understood.
© GARY BRAASCH

lakes in the high north, expanding the breeding grounds for mosquitoes, reducing
the reflectivity of the surface, and allowing increased methane production in waterlogged soils.

The characteristics of winter snow are also changing. Warmer temperatures
bring alternating freeze-thaw cycles, including rain on top of snow. In western Russia, rain-on-snow events have already increased by 50 percent during the past fifty
years. As freezing rain coats the land or snow, it forms a tough crust that is hard
for animals such as musk oxen and reindeer to break through. In Norway, Sweden,
and Finland, reindeer herding is critical to the way of life of the Saami, but these ice
crusts hinder access to lichen, a main source of nutrition for reindeer in winter. A
herder from northern Russia reports, "Nowadays snows melt earlier in the spring-
time. Lakes, rivers, and bogs freeze much later in the autumn. Reindeer herding
becomes more difficult as the ice is weak and may give way. . . . Nowadays winters
are much warmer than they used to be. Occasionally during wintertime it rains. We
never expected this; we could not be ready for this. It is very strange. . . . The cycle of

the early calendar has been disturbed greatly, and this affects the reindeer herding negatively for sure."

In the past, indigenous peoples would adjust to changes in the habitat of animals by shifting locations and moving to better ranges. This is not as easy today, in part because the previously nomadic people are settled in villages themselves, and also because the changes are too widespread. Alterations in animal migration are no longer a matter of slight shifts in timing or location. Additionally, the land is not continuous, but rather is fragmented by different uses and owners.

PERMAFROST

Permafrost occurs in soil where the maximum annual temperature is below the freezing point of water for two or more years. While the high Arctic is characterized by continuous permafrost, as one moves south toward the Arctic Circle at 66° North, the permafrost becomes discontinuous in North America and western Siberia. In eastern Siberia, continuous permafrost occurs even south of the Arctic Circle.

Permafrost also occurs in the seabed of some shallow coastal seas. Much of the subsea permafrost formed during the last ice age 20,000 years ago. Back then, lowered sea level exposed the vast Arctic shelf to atmospheric temperatures that were more than 10°C (18°F) colder than today. Sea levels were more than 120 meters (almost 400 feet) lower because so much water was tied up in the giant ice sheets that covered large parts of Canada, Scandinavia, and northern Europe.

This house collapsed due to permafrost melt near Fairbanks, Alaska. Infrastructure built on top of the previously solid permafrost is subject to tremendous subsidence as the soils melt and soften, often leading to a complete collapse, as in this case.
© ASHLEY COOPER / GLOBAL WARMING IMAGES

During summer, even regions with continuous permafrost experience summer thawing in what is called the active surface soil layer. Depending on the type of soil and the summer conditions, this layer can be as thin as half a meter, or several meters thick. Below this surface layer, permafrost extends several meters to hundreds of meters into the ground, or even more than a thousand meters in areas of Siberia. The amount of ground ice ranges from less than 1 percent to up to 90 percent, and the temperature varies from less than –10°C (14°F) to just below the freezing point. Regional differences in the character of permafrost are a consequence of both its glacial history and the present-day environment.

Measurements of permafrost temperatures around the Arctic indicate that temperatures have increased almost everywhere during the last twenty to thirty years—in some locations by just 0.5°C, but in other places by as much as 3°C (1°F to 5°F), one to six times the rate of global warming in the same period. In some regions where the decrease in snow thickness and duration will have a more significant effect than the increase in air temperatures, permafrost may actually cool. As snow depth and duration decrease, some of the winter insulation between the atmosphere and the ground is removed, allowing the ground to become colder than under thicker snow cover.

Over time, the thawing will eventually undermine large areas of the Arctic. When permafrost thaws, the frozen water drains, leaving large cavities that collapse the soil, settling it into depressions and forming uneven ground called thermokarst. Ponds and lakes form in the depressions. In other places, ponds currently found on top of frozen ground will drain.

Frozen ground behaves structurally almost like rock, but when the ice is removed, it disintegrates into looser sediments and is much easier to erode. Slopes that were frozen become unstable once the ice melts, particularly those that are not vegetated and so do not have roots to bind the soil together. In hilly and mountainous terrain, landslides and rockfalls become more frequent, especially if precipitation and storm activity increase as well.

Both ecosystems and infrastructure are affected. Plants and animals in the north have adapted to the icy conditions and have difficulty adjusting to warming. Frozen ground stabilizes the growth of trees that can tolerate some ice; when the ice is removed from around the roots, the trees settle, tilt, and sometimes keel over, leading to so-called drunken forests.

For forests built on extensive ground ice, thawing results in the formation of bogs and wetlands, which changes the range of caribou and other mammals and birds that depend on the forest habitat, and at the same time expands the range of water-loving species.

Human infrastructure is not the only thing affected by melting permafrost. Forests often grow on a thin layer of soil overlying permafrost, and as it melts, the stability of the trees is undermined. This "drunken forest" is near Fairbanks, Alaska. © ASHLEY COOPER / GLOBAL WARMING IMAGES

Over the long term, degradation of permafrost drains water from the surface layers of the soil, causing drier conditions. In Nunavat (Canada's northeastern Arctic territory), Inuit hunting and fishing grounds have already been affected, as rivers have dried up or become shallower, and the wetland habitat of swamps and bogs has diminished.

Although warming will bring extensive changes to natural habitats, it usually shifts from favoring one species to favoring another, so there are some "winners." However, human infrastructure is very closely tied to prevailing conditions. Homes, industrial and sanitation facilities, pipelines, power lines, roads, railways, and airport landing strips will all be undermined as the Arctic thaws and structures are destabilized by the shifting soil. The greatest upheaval will occur in the discontinuous permafrost zone, where perennially frozen ground is generally thinner and closer to the melting point. Structures built on hillsides may be particularly impacted by sliding and slumping of the surface layer.

Construction tailored for permafrost conditions can be used to adapt to some of these impacts. In general, most structures are designed to last thirty to fifty years, and as they reach the end of their useful life and are replaced, they can be built with changing climate in mind. By sinking stilts 15 meters or more into the ground and elevating a structure above the ground surface, building on a 1- or 2-meter-thick gravel pad, or using insulated heat pipes, structures can be kept from thawing the ground underneath them for a while, delaying the impact of general warming. However, this type of permafrost-friendly construction is expensive and its maintenance is costly.

Because of the expense, structures were often not built this way in the past. As a result, Russia in particular must now deal with widespread building deformations as well as frequent pipeline breaks and spills. One estimate concluded that almost half the buildings in Siberia were in poor condition, and damages to structures have gotten worse in the past decade.

Even where you don't see a physical structure in summer, there may be one in winter. Because summer warming of the active layer makes the tundra difficult to transit in the summer, ice roads are built seasonally in many regions of the Arctic when the surface is frozen and more stable. Transportation of goods actually increases in these regions during the winter. Just with the modest amount of warming that Alaska has experienced over the past thirty years, the ice roads across the tundra are opening later in the year—the start date shifted from October to January—and the closing date has moved from June to May. Thus, the amount of time the tundra passages are open has declined by 50 percent, from two hundred to one hundred days each year. Timber, oil, and gas operations have been impacted as their operation time has been cut in half.

Warming is causing Arctic ecosystems to relocate, shifting toward the north and expanding ranges of southern species. Cold-adapted species on land are likely to run out of room as the southern limit of their habitat reaches the coast along the edge of the Arctic Ocean. One of the major changes in vegetation is the replacement of tundra with forests. Alpine ecosystems also migrate up the flanks of mountains. In northern Sweden, tree lines have already moved as much as 60 meters (almost 200 feet) up the mountainsides during the last century. These changes darken the landscape, further enhancing the warming, because trees, especially conifers, tend to be darker than the low-lying shrubs, grasses, mosses, and lichens of the tundra.

Changes in climate and vegetation also result in an increase in fires, which are enhanced under warmer and drier conditions, as is the availability of flammable materials and ignition sources, including those from human activity. Along the southern periphery of northern forests, weakening of trees by insect infestations, permafrost degradation, and land-use changes provide fuel for fires. At the same time, expanding the range of highly flammable black spruce into northern tundra habitat likely increases fire frequency and intensity. The area of northern forest

Forest fires, such as this one in Alaska, are on the increase in the Arctic due to drier summers, insect infestation, and increases in human activity. © PETER ESSICK

Arctic soils are generally held together by ice and permafrost. As the permafrost melts, the soils become very sandy and easy to erode. Along the coast, sea ice retreat also exposes the soils to increased ocean-wave activity. When both effects combine, as at the town of Shishmaref on the Bering Strait, coastal erosion rates can reach more than 20 meters (60 feet) a year. In the case of Shishmaref, this location is being completely abandoned.
© GARY BRAASCH

burned annually in Russia more than doubled in the 1990s, and there have been increases in Alaska as well.

There are also global consequences of Arctic permafrost thawing: bacterial decomposition of organic matter in these northern soils has the potential to release greenhouse gases to the atmosphere, potentially exacerbating future warming. If the decomposition occurs in air, it produces carbon dioxide, but when oxygen is not present, as in swamps and waterlogged ground, decomposition produces methane, a shorter-lived but more powerful greenhouse gas. The amount of organic carbon stored in just the upper 25 meters (80 feet) of the Arctic permafrost is huge—750 to 950 gigatons of carbon (10^9 metric tons). This is comparable to the 800 gigatons in the atmosphere today.

Thawing permafrost combined with sea ice changes are causing major disruptions along Arctic coasts. In the past, sea ice would form in early fall along the shore, protecting the land from the full force of fall and winter storms. Now, as the sea ice armoring occurs later or not at all, high waves and storm surges are eating away at the coasts and carving into shorelines.

Much of the Arctic coastline consists of sand, silt, and gravel deposits. Because permafrost helped bind these sediments together in the past, they were resistant to erosion. Now, as the permafrost thaws, the soil is disaggregated and sediments are easy to wash away. Rising sea levels also contribute to the problem. Coastal erosion rates along the Beaufort Sea in Alaska have more than doubled in the last fifty years.

Many coastal communities—some inhabited for thousands of years—are now being forced to consider evacuation. As sea ice has formed later along the coasts and has become more unstable, coastal communities have lost their sea ice roads. These passageways were critical for winter travel to hunting grounds, for island communities to hunt on the mainland, and for coastal communities to hunt at sea.

It is not just the coastal communities that are being affected. Ports, including industrial facilities such as tanker terminals, and offshore structures also are threatened by rising sea level, receding shorelines, and less stable sea ice conditions.

In Alaskan lakes, bacterial decomposition of organic matter continues even when the lakes are frozen over. These bacteria give off methane gas, which collects under the icy surface. By breaking a small hole in the ice, the gas can be released and lit by intrepid researchers, in this case by Katey Walter of the University of Alaska. © MARMIAN GRIMES / UNIVERSITY OF ALASKA FAIRBANKS

CHANGING ECOSYSTEMS

Changing ecosystems in the sea also affect ecosystems on land. Consider the marginal ice zone, where the drifting pack ice encounters open water. This border is a region of high productivity during the spring; in fact, the productivity is so high that birds and whales migrate thousands of kilometers from all over the globe to feed here. And we do, too: Arctic fisheries, including the Barents and the Bering seas, are among the most productive in the world, providing about 10 percent of the world catch. Alaskan fisheries alone provide more than half of the fish caught in U.S. waters.

The marginal ice zone is an area of high productivity because nutrients, built up in surface waters under the ice during the long winter, are finally exposed to sunlight in spring. The combination of sunlight and nutrients culminates in a burst of algae blooming in the surface waters from April through May. The algae are eaten by zooplankton, which in turn are fed on by fish and seabirds. Biological activity during the spring bloom is so intense because much of the food is kept floating in the upper part of the water column due to the density stratification caused by

CLIMATE CHANGE

melting ice. The freshwater melt makes the surface layer less dense than the more salty water beneath. It is like adding pepper and other spices to an oil and vinegar salad dressing: the spices settle through the top layer of oil and then remain there at the interface along the top of the vinegar. But because so much food is produced in the ocean during the spring bloom, not all of it can be eaten right away, and some settles to the sea floor. In shallow regions, this rain of food supports bottom-dwelling organisms such as clams, crustaceans, marine worms, and other species that feed from the sea floor, including bearded and ringed seals, walruses, gray whales, and eider ducks.

The gray whales of coastal California come all the way up to the Bering Sea between Alaska and Siberia to feed off crustaceans living on the sea floor: a 10,000 kilometer trip that is one of the longest migrations made by a mammal on an annual basis. These temperate whales return to southern climates for the winter to bear their young. In contrast, the Arctic whales—narwhals, with their magnificent unicorn-like twisted horn, and white and bowhead whales—have much shorter migration patterns, as they follow the edge of the pack ice north when it retreats in summer and south as it extends in the winter. Whales in and around ice-covered waters need to find open spaces between floes, or along their edges, where they can come to the surface to breathe. The bowhead whale, however, can break through thin ice cover with its head to make a small breathing hole.

Like the Arctic whales, many seabirds are local, while other migrate in from far away for the summer to feed and breed. The Arctic tern is the champion migrator: it summers in the Arctic for breeding and then flies down to the Antarctic to catch the Southern Hemisphere summer as well. In the Arctic, seabirds nest on land, along cliffs (to escape predators), or on the ground, but they fly out to sea to forage along the edges of the sea ice and in other places where prey is concentrated by ocean currents.

As the extent of sea ice changes, so will the location of the marginal ice zone. More open water will benefit some species, especially those that already exist in temperate zones, such as harbor and grey seals. Unfortunately, the open water disrupts the feeding and breeding patterns for others. For example, when the marginal ice zone is too far from land, birds that nest on cliffs are not able to combine their feeding with breeding.

In recent years, patchier and thinner sea ice coupled with rising air and water temperatures are shifting the northern Bering Sea habitat toward conditions usually found farther south. The native fish and other animals that the local Inuit depend on for their subsistence and lifestyle are getting harder to find. Migrating gray whales are ranging farther north, raising concerns that they will begin to com-

A coccolithophore bloom in the Barents Sea (north of Norway) seen through a break in the clouds in August 2007. Coccolithophores are tiny surface-dwelling plankton that form calcite (limestone) scales that can clearly be seen in the water. In the normally nutrient-poor subpolar waters, storms or seasonally retreating sea ice can bring nutrients to the surface, setting off a rapid increase in plankton growth (a bloom). NASA IMAGE COURTESY OF NORMAN KURING, OCEAN COLOR GROUP, GODDARD SPACE FLIGHT CENTER

pete with local Arctic species. Also, as the marginal ice zone retracts even farther to the north, away from the shallows around the continents, all the excess food produced during the spring bloom will fall into the deep ocean abyss. Waters in the central Arctic basin are thousands of meters deep rather than the tens or hundreds of meters deep around the periphery. Gray whales are not able to dive this deep and so will lose their major food source.

> **Being ever prepared, "upterrlainarluta," is a common caution from Yup'ik elders to young people, whether they are preparing for fishing or a trip to the city. Implicit is the understanding that one must be wise in knowing what to prepare for and equally wise in being prepared for the unknowable.**
>
> —James Barker, *Always Getting Ready*

Climate change is, however, not the only environmental stress on the Arctic—stratospheric ozone loss and the transport of other pollutants from lower latitudes are also concerns. While stratospheric ozone loss is not as dramatic as in the Antarctic, it is also a problem in the northern latitudes. There is a seasonal cycle to ozone loss, and the lowest levels occur during spring, just when the Arctic ecosystem is blooming after the long winter. Ozone loss allows more ultraviolet radiation to reach the Earth's surface, causing damage to eyes and skin as well as to the immune system and cell growth. International policies to reduce the production and use of ozone-depleting chemicals are working to help the ozone layer recover, but progress is slow: anomalous ultraviolet radiation will continue to be a concern in the Arctic for decades to come.

Despite the remoteness of the Arctic and its seemingly pristine condition, agricultural and industrial pollutants are transported north with the same winds that bring heat from the tropics. Once there, the contaminants move through the physical and biological system. Because some chemicals, such as PCBs and insecticides, attach to fat and do not degrade easily, they wind up accumulating at the top of the food chain—and you are what you eat. As a result, polar bears in Greenland and other regions of the Arctic have such high levels of contaminants in their bodies that their reproductive ability may be impaired. Combining this toxic load with difficulties in obtaining prey due to climate change may be the downfall of some higher-level Arctic organisms, including seabirds and polar bears: as animals starve, they draw on their reserves of fat, and if it is loaded with contaminants, this will exacerbate health problems. Eating contaminated organisms also exposes Arctic people to health risks: blood samples from a number of residents of eastern

Canada, Greenland, and eastern Siberia have indicated elevated levels of persistent organic pollutants such as PCBs and the insecticide DDT.

Other pressures impinging on the Arctic from the south include a demand for fish, oil, natural gas, and other resources, as well as agricultural development and an increase in tourism. As the Arctic people and the ecosystem struggle to adapt to change, new economic and governance issues are arising. The extended open season of northern passageways is already resulting in interesting geopolitics: the Northern Sea Route is governed by Russia and the Northwest Passage by Canada, though this is challenged by the United States. The planting of the Russian flag on the sea floor at the North Pole and Canada's decision to open new Arctic bases in 2007 are just the beginning of national jockeying for position. Siberia is rich in resources—coal, oil, natural gas, timber, and minerals such as copper and nickel—that will become more accessible as the sea route grows cheaper and more reliable (at the same time thawing permafrost further degrades transport routes over land). Shipping, offshore development, commercial fishing, and tourism will exert pressure on regions that used to be available solely to local residents and migrating species. Who will decide on environmental protocols in newly opened areas? The question applies not just to these major shipping routes, but to passages all around the Arctic perimeter.

MEANWHILE, AT THE OTHER END OF THE PLANET...

Antarctica is almost entirely covered by a giant ice sheet that is nine times the volume of that covering Greenland. With ice thicknesses ranging up to 5,000 meters (16,400 feet) and an average thickness of about 2,400 meters, Antarctica is the highest and coldest continent, as well as the driest and windiest. Because it is so cold, even summer temperatures rarely go above freezing, and the ice sheet does not lose much ice by melting. The ice sheet is normally kept in balance as snow added to the interior compensates for the ice flowing to the periphery and into the ocean, where it breaks off and forms spectacular icebergs. Much of the Antarctic glacier ice is fed into a floating ice shelf somewhere along its coast.

Geographically, Antarctica can be divided into three areas: the Antarctic Peninsula, the horn of land that extends north toward the southern tip of South America; the West Antarctic Ice Sheet, the sector to the south of the peninsula; and the much larger East Antarctic Ice Sheet, located on the other side of the continent and including the South Pole.

Some of the most dramatic changes have been observed on the Antarctic Peninsula. In 2002 there was a stunningly rapid loss of ice from the Larsen B ice shelf

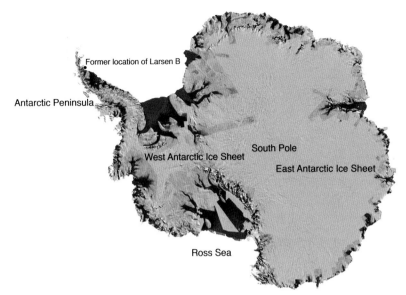

Former location of Larsen B

Antarctic Peninsula

West Antarctic Ice Sheet

South Pole

East Antarctic Ice Sheet

Ross Sea

Antarctica is a vast and vastly underexplored continent, so the most comprehensive data often come from satellites. In this image, data from a satellite interferometer, which measures how fast the ground is moving from subtle changes to the reflected light, have been used to identify the fastest-moving glaciers around the edge of the continent. Green represents slow-moving ice, and purple denotes much faster flows. Glaciers on West Antarctica and the Antarctic Peninsula (to the left in this image) are draining the ice faster than snowfall is replenishing it, thus adding to overall sea level rise. NASA

on the eastern (Atlantic) side of the peninsula. Intense atmospheric warming in the region—more than 2°C (4°F) within fifty years—had caused melt ponds to form on the surface of the floating ice shelf. The meltwater had penetrated into ice crevasses, helping to wedge the shelf apart. At the same time, the ocean that the ice was floating on had warmed. This thinned and stressed the ice shelf, literally to the breaking point. Within a period of thirty-five days, 40 percent of the ice shelf, a region the size of the state of Rhode Island, disintegrated into thousands of icebergs—the largest such event during thirty years of monitoring. Subsequently, the land-based glaciers that fed ice into the shelf were no longer buttressed by ice pressure from merging glaciers, and so their flow accelerated, reaching speeds up to eight times their previous value. This pattern of ice shelf breakup, followed by draining of ice from the interior, is something that raises even greater concerns with respect to the huge West Antarctic Ice Sheet just to the south.

The West Antarctic Ice Sheet is about 10 percent of Antarctica's total glacial volume (5 or 6 meters [16 to 20 feet] sea level equivalent) and feeds ice into the continent's two largest ice shelves: the Ross ice shelf in the South Pacific, and the Filchner/Ronne ice shelf in the South Atlantic. Each of these ice shelves is the size of Spain. This, in and of itself, means the interior ice is not as stable as it would be if the glacier terminated on land. But another factor is at play here: the land that the West Antarctic Ice Sheet covers actually sits, on average, almost 500 meters (1,600 feet) below sea level. This means that it is possible for ocean water to penetrate beneath the ice sheet, lubricating the base and accelerating its flow into the sea.

Measurements from satellites show that ice shelves in the Pacific sector of the West Antarctic Ice Sheet have been thinning by as much as 5.5 meters (18 feet) per year over the past ten years. The thinning probably is caused both by the warming of ocean waters in the region (by about 0.5°C [1°F]), as well as by an increase in the flow speed of the glaciers feeding into the shelves. Because the interaction of the ice shelves with atmospheric and oceanic warming and the basal water regime is so complex, the future behavior of the West Antarctic Ice Sheet is one of the most uncertain components of Earth's climate system.

Another Antarctic uncertainty, at least for the short term, is how the loss of ice

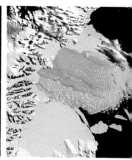

along the periphery of the continent may be balanced by increased deposition of snow in the interior. Antarctica is classified climatically as a desert, with average annual snow accumulation rates in the interior of 5 centimeters (2 inches) water equivalent per year. Snowfall is expected to increase in the future as warmer air transports more moisture to the region. This increase in snowfall has, so far, offset some of the sea level rise caused by loss of ice along the edges. Recent results from satellites measuring the gravity anomalies of the planet indicate that the Antarctic is currently losing mass, but just fractionally, and the long-term budget is still uncertain.

Temperature change data on the continent are noisy and incomplete due to the sparseness of the coverage. Southern Hemisphere temperatures in general are warming more slowly than in the Northern Hemisphere—an expected result, because the larger area of open ocean in the south leads to greater thermal inertia. Southern sea ice is more stable as a result and has not yet seen any significant trends in recent years. Some things have changed noticeably, though: winds that circle the continent have increased in speed, possibly as a function of the changes in atmospheric conditions due to the polar ozone hole; this change in turn affects temperatures, particularly on the peninsula, and the rate of sequestration of both carbon and heat in the deep southern oceans.

A GLOBAL CONNECTION

Each of the components of the polar fabric has a global connection. The extent of sea ice affects planetary reflection of sunlight, the circulation of the world ocean, and the global transport of goods. The storage of water in glaciers affects sea level worldwide; the carbon stored in permafrost or sequestered in the deep ocean affects the global concentration of greenhouse gases; and resources such as fisheries, timber, minerals, and fuel sustain populations and development around the world. Through these connections, what happens there affects us all.

The Larsen B ice shelf on the Antarctic Peninsula collapsed in a matter of weeks from January to March 2002. The first image shows the ice shelf with a few unusual surface melt ponds. The next two images show small collapses spilling ice into the ocean. The fourth image shows the full collapse, a matter of days after the previous image was retrieved. The final image shows the icebergs from the shelf spread over a wide area, turning the ocean a bluish white with icy debris. TED SCAMBOS, NATIONAL SNOW AND ICE DATA CENTER

Elizabeth Kolbert

REPORTING ON CLIMATE CHANGE

A few years ago, I set out to visit places changed by global warming. The first place I traveled to, via Nome, was an Inupiat village named Shishmaref, which sits on a small island in the Chukchi Sea, five miles off the coast of the Seward Peninsula. I'm a reporter, and I went to Shishmaref for the reason reporters usually go places: because something was happening there. The village used to be protected from fall storms by the sea ice, which began to form in mid-autumn. Now that the sea ice is forming later and later, the island is, quite literally, disappearing; every few years, another swath falls into the sea. At the time I visited, the situation in Shishmaref was so dire that residents had voted to move the entire town to the mainland. By telling their story, I hoped to convey the "now-ness" of global warming. Its effects will be felt not just in some hypothetical future; they can already be seen in the present.

But here's the paradox: whatever you can *see* is not actually the global warming of the present. The effect of pumping greenhouse gases into the air is to throw the planet out of energy balance; more energy is being received from the Sun than the Earth is radiating back out to space. For balance to be restored—as, according to the laws of physics, eventually it must be—the entire Earth has to warm up. This is a slow process; the Earth is a big place, and a lot of it's covered by oceans. Thus, the warming that we are seeing today—and that is driving Shishmaref into the sea—is a function of greenhouse gases emitted decades ago, and the full effects of the carbon dioxide we are emitting today will not be felt until decades from now. Global warming in this way is always much farther along than it appears. Another way to put this is that whatever you can see is the past of climate change; the present is still invisible.

Meanwhile, even if one could see global warming in real time, could one say for sure that what one was seeing was global warming? After Hurricane Katrina, many Americans wanted to know whether New Orleans had become the first major U.S. city to fall victim to climate change. The correct answer, scientists pointed out, is that this is the wrong question. The fact that climbing carbon dioxide levels are expected to produce more storms like Katrina doesn't mean that Katrina was caused by global warming. No single storm—or flood or drought or heat wave—can ever be so conclusively attributed; weather events are a function both of factors that can be identified and of factors that are stochastic, or purely random. This is another aspect of the problem that bedevils coverage. News is a form of narrative, and narrative demands particulars—*this* storm, *this* city, *these* people. Rising sea surface or average global temperatures, by contrast, are abstractions, things that no one in particular experiences.

When journalists aren't writing about events, usually they are writing about conflict: Democrats versus Republicans, attorneys for the prosecution versus attorneys for the defense. Reporters are supposed to stand outside the conflict and treat both sides impartially. But, as many have noted, this effort to be unbiased can itself become a form of bias. Just because two people, or two groups, disagree, it doesn't follow that the two sides have an equal claim on our attention, or that there's some middle position that better reflects the truth of the situation. The reporter's habit of giving equal time to the opposing sides on any issue is easily exploited, and of course has been by so-called global warming skeptics. People with no particular knowledge, supported by groups with obvious vested interests, have often been accorded just as much column space or air time as scientists who have devoted their lives to studying the issue. Journalists, it could be argued, completely missed the global warming story by treating it as a debate when, really, it never was one.

It is still difficult to write about global warming, although, sadly, it is becoming less so every record-breaking year. As the world changes, will the coverage manage to convey what is truly at stake? Obviously one hopes so. But journalism will always favor the here and now, along with easy-to-grasp causalities and dramatic confrontation. Even as the old debates are being overtaken by events, new debates will arise and may prove equally distracting. (Are the costs of combating climate change worth it? Is climate change, for some, perhaps a good thing?) For better or (mostly) for worse, writing about climate change has a long future ahead of it.

CHAPTER 3

SEA CHANGES

Anastasia Romanou

And thou, vast ocean! on whose awful face
Time's iron feet can print no ruin-trace.
                                        —Robert Montgomery

Ocean: A body of water occupying two-thirds of a world made for
man—who has no gills.
                        —Ambrose Bierce, entry from *The Devil's Dictionary*

On January 10, 1992, some 29,000 brightly colored plastic ducks (and turtles, beavers, and frogs) were accidentally thrown off a ship in the North Pacific Ocean near the international date line and started a journey that has taken some of them tens of thousands of kilometers. By September of that year, these plastic toys started washing up along the shoreline of Alaska. Toys showed up in Hawaii in 1996, and by 2003 at least one duck and a frog had made it through the Arctic Ocean to the North Atlantic, landing as far south as Maine and Scotland.

The paths taken by these floating toys illustrate the great ocean-gyre current systems, such as the subarctic gyre in the North Pacific, or the East Greenland Current and North Atlantic Drift. Just as these currents on the surface of the oceans transport the toy ducks, they transport heat, salt, and dissolved nutrients across the world's oceans. The currents are driven mainly by the prevailing winds that tug at the ocean's surface. Near the equator, the prevailing winds blow toward the west. Above and below the equator in the midlatitudes, they blow eastward. The resulting currents form basin-wide gyres that flow up to 10 kilometers (6 miles) a day. In the Northern Hemisphere they go clockwise in the subtropics and counterclockwise in the subarctic (and the other way round in the Southern Hemisphere). For instance, the round trip of between 7,000 and 14,000 kilometers around the North Pacific

A selection of plastic toys from Curtis Ebbesmeyer's collection of beach-combed objects that he has been using to track ocean currents. © JOSHUA WOLFE

73

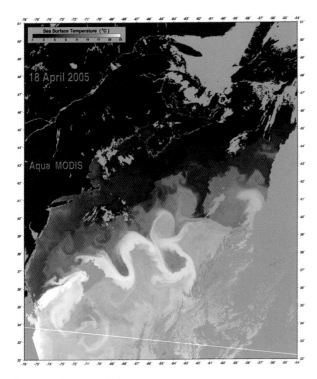

Sea Surface Temperature (°C)

18 April 2005

Aqua MODIS

The meanders of the Gulf Stream once it separates from the U.S. coast at Cape Hatteras produce warm- and cold-water rings that can persist in the Atlantic for months. NASA IMAGES COURTESY OF NORMAN KURING, MODIS OCEAN TEAM

gyre takes two to four years, and almost like clockwork, some toys have appeared according to schedule on the beaches of Sitka, Alaska, for at least four of these rotations.

The density of ocean water is a second factor that drives circulation within the ocean. The waters in the oceans vary a great deal in temperature and content of dissolved minerals—their salt content (or salinity) and their density depends quite strongly on this composition. Cooler or saltier waters are more dense (heavier) and will therefore sink relative to a warmer or fresher layer of water. (Think of the different layers as oil and vinegar in a salad dressing—if left to settle, the oil, which is lighter, will always float above the vinegar.)

Processes that affect water density therefore drive changes in circulation. Rainfall, ice melt, and river outflow are all sources of freshwater that tend to dilute the ocean surface water, making it less dense. Evaporation does the opposite, increasing the salt concentration of the water left behind. Similarly, cooling of the ocean by the wind or at night makes the surface water denser, whereas heating from the Sun makes it more buoyant. The mixing of ocean water by the tides, winds, or convection also affects water density and, hence, ocean circulation.

In the polar regions, the ocean is cold and often covered with sea ice. As this relatively fresh ice forms at the ocean's surface, it pushes out much of the salt, leaving the waters below the ice more salty than before. The resulting very dense, cold seawater brine sinks, often to great depths. This deep ocean convection takes place in specific areas, such as the Labrador and the Greenland seas in the North Atlantic and the Weddell and the Ross seas around Antarctica, during episodic events each winter. Sinking cold plumes from deep currents follow the contours of the sea floor and flow toward the equator, while slowly mixing with surrounding waters. This flow is called the ocean thermohaline circulation, and is named for the role of temperature and salt in creating the deep waters (*thermo* means heat, and *haline* means salt). Through tidal and turbulent mixing in the ocean interior, the water in this deep circulation eventually makes its way back to the surface, most noticeably in upwelling zones in equatorial regions near the western coasts of the continents. Such protrusions of cold water of deep origin (named cold tongues) can be seen clearly in

temperature maps of the eastern Pacific and the eastern Atlantic oceans. If one of the lost toy ducks were able to follow these underwater currents, it would spend centuries or even millennia in the deep ocean before it came back to the surface.

This vast and slow-moving deep circulation gives the ocean an enormous inertia. It plays a dramatic role in how the ocean absorbs and sequesters heat and atmospheric pollutants. It is only at these few deep convection sites that the ocean can absorb greenhouse gases from the atmosphere—principally carbon dioxide (CO_2)—and store them away in its deeper layers (the mixing of heat and gases elsewhere in the ocean is greatly limited by the strong contrasts between the buoyant surface waters and the denser deep waters and is not at all efficient).

The enormous heat capacity of the ocean means that changes in the atmosphere take a long time to affect the ocean. Yet over the last fifty years, the oceans have been warming at almost all levels, and in fact, the amount of extra heat energy that has been accumulated in the oceans is twenty times the amount of extra energy in the atmosphere or land surface. However, sea surface temperatures have increased only slightly less quickly than surface air temperatures.

Poetically, the ocean is often referred to as unchanging and unchangeable, but

A view from the *Nathaniel B. Palmer*, one of two U.S. icebreakers operating in Antarctica that provide a platform for studies of the ocean, sea ice, and glaciers around the continental margin. It is named after the nineteenth-century captain who first sighted the Antarctic Peninsula in 1820 (the southern part of which is now called Palmer Land). © GARY BRAASCH

this simply isn't the case in the real world. The changes can be major, they can even be rapid, and they already are playing a large role in ongoing climate change.

El Niño events are a great example of dramatic, natural, ocean changes. The El Niño/Southern Oscillation is a natural phenomenon that occurs every two to seven years in the equatorial Pacific Ocean and lasts for a few seasons. It has been known for centuries, particularly to the Chilean fishermen who associated it with reduced offshore catches. It most often arrives around Christmas, and they therefore named it "the Child." El Niño can be triggered when the normally westward-blowing equatorial trade winds weaken or even reverse direction as part of the normal variation in tropical weather. This change in the winds allows a large mass of warm water from the West Pacific Warm Pool, normally situated near Indonesia, to extend eastward along the equator until it reaches the coasts of South America, where the cold tongue usually exists. During El Niño, the wind patterns that push water away from the coast and bring up cold, nutrient-rich water from the deep are shut down (hence the impact on the fisheries). El Niño is typically followed by the opposite phase in the cycle, with colder-than-usual surface waters over the tropical Pacific. Scientists have tried out a number of names for this opposite phase—El Viejo ("the old one") or the anti–El Niño (though given the religious connection, this was quickly dropped)—but the one that stuck is La Niña ("the little girl").

An El Niño event can have devastating impacts on tropical Pacific rainfall—causing flooding in South America and droughts in Indonesia and Australia—and have noticeable effects as far afield as California and southern Africa. In particular, historical droughts in the American Southwest are associated with La Niña conditions, whereas increases in California rainfall are sometimes associated with El Niño. These correlations are based on changes to upper-level wind patterns, but though the odds change in El Niño years, these far-field responses aren't 100-percent reliable. In the Atlantic, El Niño years are associated with strong reductions in hurricane activity, again because of atmospheric wind changes.

SEA LEVEL

As mentioned earlier, ocean circulation plays an important role in sea level. Where there are colder surface temperatures the ocean surface bends in like a bowl, and where there are warm waters the surface of the ocean is a little higher than its surroundings. In the center of the subtropical gyres (for instance, to the east of the Gulf Stream), the sea level stands about a meter higher than at the coast.

Other natural factors also influence sea level. Much like the land surface, the sea surface also has its "topography" of hills and troughs, valley and ranges—even

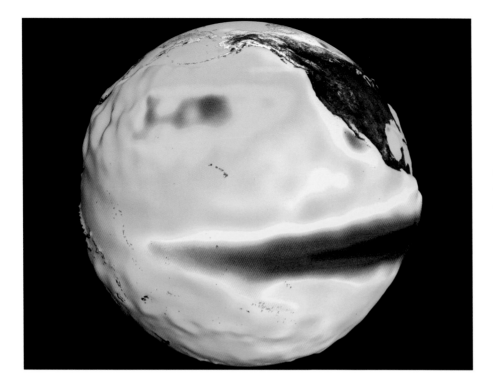

The sea surface temperature changes associated with the 1997–1998 El Niño event shown with the associated sea surface height changes (highly exaggerated!). In the equatorial eastern Pacific, the sea level is higher because the water expands as it warms.
NASA/GODDARD SPACE FLIGHT CENTER SCIENTIFIC VISUALIZATION STUDIO

if one were to smooth over all the surface waves. Most features are about a few centimeters in height and are the result of anomalies in the Earth's gravitational field; bumps on the surface tend to follow such ocean-bottom features as seamounts and trenches.

The lunar and solar tides are the most visible signs of sea level variation, but natural events such as El Niño and La Niña also cause changes to sea level height. For instance, during the 1997–1998 El Niño, sea level on the west coast of the Americas increased by over 30 centimeters (1 foot), while in Indonesia it was lower than normal. Individual storms can change local sea level as well; the storm surge driven by the winds from a land-falling hurricane or Nor'easter can be measured in meters, and it is usually the most damaging feature of the storm.

The aspect of sea level change that is of most concern in climate change is not related to these local or regional effects, but to what is described as eustatic (global) sea level rise. There are two main pieces to the global numbers: First, in a warming climate, sea levels rise simply because ocean water increases in volume when heated (thermal expansion), and as long as the ocean continues to warm, the rise will continue. Given the long timescales for reaching the deep ocean, this fact implies that sea levels will continue to rise for centuries after atmospheric temperatures have stabilized. Second, the total amount of water in the ocean can change, mostly from the melting of ice on land and the extraction of groundwater. Note that the melting of ice that is already floating (such as Arctic sea ice or the great Antarctic ice shelves) only makes a very minor contribution to sea level rise. This ice is already displacing seawater, and the only reason the contribution is positive at all is because the melted ice is less salty and so has a slighter lower density (or greater volume). Thus, the direct impact of floating ice is usually neglected in sea level assessments; however, see Chapter 2 for examples of indirect connections between ice shelf changes and increased calving of the land ice that feeds them.

Since the nineteenth century, sea level rise has increased from about 0.5 millimeter per year to at least 2 millimeters per year, and possibly over 3 millimeters per year in recent decades. Global sea level has already risen by 10 to 20 centimeters (4 to 8 inches) in the past century, leading to noticeably more coastal flooding events. Since 1993, thermal expansion has accounted for about half of the rise, with melting of mountain glaciers, ice caps, and ice sheets (Greenland and Antarctica) accounting for the rest. As discussed in Chapter 2, Greenland is almost certainly losing ice mass, while Antarctica is closer to mass balance since losses in West Antarctica and around the periphery are approximately being offset by snow accumulation in the interior. The uncertainty in future sea level trends depends very strongly on what will happen to those ice sheets.

OCEAN CHEMISTRY

The ocean has rightly been called a chemical soup, as it contains huge numbers of chemical compounds, elements, gases, minerals, and organic and particulate matter. Climate change is already impacting the chemical consistency of the ocean by changing the distribution of salt and nutrients, the carbon content, and the acidity of seawater.

As described earlier, when ice forms directly from seawater, the salty brine cannot be included in the ice crystals and is squeezed out, leaving behind predominantly freshwater ice. But when this ice melts, freshwater returns to the ocean; and the more it melts, the more the ocean freshens. As Arctic sea ice has reduced in extent over the last thirty years and midlatitude rainfall has increased, a dramatic, albeit gradual, freshening of the North Atlantic has taken place. Freshening makes the water lighter, and unless it is accompanied by some cooling (which has the opposite effect), freshening makes deep convection more difficult. (Deep convection events, as mentioned earlier, provide the pathways for greenhouse gases that are absorbed from the atmosphere to be transported and stored in the deep ocean.) Conversely, increased water use for irrigation and damming of rivers has led to increased salinity in some select coastal regions. Such significant and persistent alterations in the salinity of the ocean not only affect density and circulation, but also affect the biodiversity in marine ecosystems, as we will see later.

The oceans more generally are the repository (or "sink") for much of what gets dumped, discarded, and exuded in rivers and the air. Roughly one-third of all the CO_2 produced from fossil fuel burning and other human activities has ended up in the ocean, mostly through gas exchange at the surface. The details of this process are complicated by the fact that the ocean is what is called a buffered solution. Buffering occurs when CO_2 reacts with seawater and produces bicarbonate and carbonate ions (HCO_3^- and CO_3^{2-})—the balance between these compounds depends on the acidity (pH), temperature, and salinity of the water. Although adding more CO_2 increases the acidity of the water, buffering causes the ocean to resist that change, thus making the uptake of excess CO_2 much slower than for more inert gases. Incidentally, it was the discovery of this buffering by Roger Revelle and Hans Suess in the 1950s that led to the realization that industrial emissions of CO_2 could in fact pose a climate threat because the ocean wouldn't immediately take up the excess.

The anthropogenic (human-induced) increase in dissolved CO_2 already has led to a decrease in ocean pH of about 0.1 unit (from 8.2 to 8.1) since the Industrial Revolution, and even though the ocean technically is becoming less alkaline (acids are defined as having a pH of less than 7), this change has been termed ocean acidification. As the ocean continues to absorb excess CO_2, it is estimated that the pH

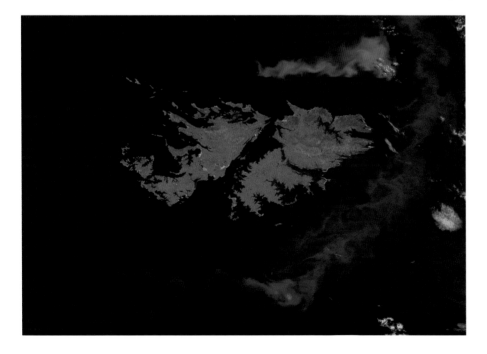

A phytoplankton bloom surrounding the Falkland Islands in the South Atlantic during the Southern Hemisphere summer in January 2008.
IMAGE COURTESY OF JEFF SCHMALTZ, MODIS LAND RAPID RESPONSE TEAM, NASA/GODDARD SPACE FLIGHT CENTER

Australia's Great Barrier Reef is the largest coral reef system in the world, and is home to thousands of species of fish, mammals, crustaceans, and other sea life. PROVIDED BY THE SEAWIFS PROJECT, NASA/GODDARD SPACE FLIGHT CENTER, AND ORBIMAGE

sturdy home for the algae. In concert, these organisms build the enormous coral reef ecosystems that are legendary for their productivity and biodiversity.

With large changes in the temperature or the chemical composition of the surrounding waters, corals are known to expel these algae, losing their signature bright color and turning to light yellow or even white, a stress reaction known as coral bleaching.

The Great Barrier Reef along the northeastern coast of Australia suffered two mass coral bleaching events in the summers of 1998 and 2002, associated with exceptionally warm temperatures in those years. Although most reef areas recovered eventually with little coral loss, some locations suffered severe damage, with up to 90 percent of the corals killed. Extensive coral communities in the Indian Ocean have been damaged permanently by warm sea temperatures, with up to 90 percent of corals lost in the Maldives, Sri Lanka, and the Seychelles. In 1998, 75 to 99 percent of corals were lost to bleaching in most countries of the western Indian Ocean. In 2005, an anomalous warm spell caused bleaching of 52 to 85 percent of the coral cover in the Caribbean.

The difference in species diversity between piles of dead coral (seen here near Gan in the Maldives) and a thriving reef ecosystem (seen here on House Reef, Cabilao, Philippines) is profound and has implications far beyond the coral itself. ABOVE: DAVID R. BURDICK; OPPOSITE: © PETER ESSICK

When corals die in such massive numbers, the entire ecosystem can collapse. Bleaching leads to loss of reef complexity and undermines the livelihood of other reef communities that depend on the corals for shelter and food. Some invertebrate and fish species forage on coral larvae. Others use the reefs for refuge and protection from larger predators. Many fish in turn provide nutrients to the coral organisms. In coral reef communities where intense bleaching has occurred, significant decreases in the abundance of other fish are observed as well.

At the same time, ocean acidification lowers the concentration of carbonate in the water, making it difficult to produce calcium carbonate, the building material for coral reefs. Organisms grow more slowly and their skeletons become less dense, very much like osteoporosis in humans. As a result, reef structures are threatened because corals may be unable to build reefs as fast as erosion wears them away. Decalcification threatens other marine calcifying organisms as well, such as marine plankton snails (pteropods), marine plankton algae (coccolithophores),

echinoderms (starfish and sea urchins), crustaceans (lobsters, crabs, shrimp, crayfish, and barnacles), and some mollusks (snails and clams), which also use calcium carbonate to build their skeletons and shell coverings.

It is difficult to know exactly to what extent climate change is responsible for widespread coral die-offs. Bleaching due to warming and destruction due to increasing storm damage are happening at the same time as overfishing, the spread of other coral diseases, pollution, and habitat destruction. But climate change, at minimum, exacerbates an already serious problem.

FISHERIES

Along the Pacific coast of the Americas, the impacts of El Niño illustrate clearly how climate can directly affect marine fisheries over short periods. For example, higher sea surface temperatures associated with the 1997–1998 El Niño resulted in a collapse of California's squid fishery as the animals moved to cooler, deeper waters to spawn. Landings fell by more than 95 percent compared to the record-breaking 1996–1997 season, which was not an El Niño year. During the same El Niño event, widespread sea lion pup deaths occurred in California due to the significant reduction in their food source. Warm-water marlins migrated and were caught as far north as Washington State. Cold-water salmon almost disappeared from Bristol Bay in Alaska.

Similarly, in a changing climate, long-term increases in temperature and changes in ocean circulation, salinity, light, and nutrient distribution are expected to affect species populations, their migratory patterns, and their primary habitats. These disruptions will cascade up the food chain.

Most fish can adjust their body temperature and therefore increase the efficiency with which they turn food into energy in order to adapt to their environment. Others grow faster when it gets warmer, like the northern Pacific Coast English sole populations, which grow heavier and larger as water temperatures rise. Similarly, several studies have shown that the growth rates of some coastal and long-lived deep ocean fish, such as morwong, redfish, and orange roughy in the southwest Pacific Ocean, also have increased due to warming temperatures.

When marine organisms cannot change their growth and metabolic rates, they will respond to their changing environment with migration, relocating to areas with more favorable conditions. For example, in the northern Pacific Aleutian region, killer whales had to drop sea lions and seals from their diet when the latter changed habitats due to increasing temperatures. Whales then turned to sea otters, causing a substantial decline in otter numbers. The otters in turn ate fewer sea urchins, whose populations expanded to the detriment of kelp beds (their food) and the diverse communities that lived there.

In the North Sea, changes in wind patterns and ocean circulation have led to warmer temperatures, which have caused phytoplankton to migrate northward. Animals that depend on the phytoplankton, such as zooplankton and larger fish like mackerel, followed their food farther north, greatly reducing mackerel catches in their traditional area.

Polar fish have a narrower tolerance to environmental conditions than fish in other regions, so as waters change temperature, they often seek deeper or shallower waters that are more comfortable. For instance, in the 1980s, following particularly strong convective events in the Weddell Sea a decade earlier, the Antarctic bottom waters in the Argentine basin cooled by about 0.05°C and freshened by 0.008 part per thousand, which was enough to drive polar fish away. The food chains in the polar regions are short and depend on relatively few species. In the Southern Ocean, plankton are trapped by the Antarctic convergence, which restricts their passage to the north.

Cod catches in the North Atlantic fell sharply in recent decades, and overfishing was the principal culprit. However, cod is near the top of the food chain, eating a variety of other species, and changes in phytoplankton distribution, possibly caused by climate change, may have been a contributing factor in the reduction of Atlantic cod stocks and their failure to recover once fishing was restricted. Changes in where fish live and how numerous they are eventually affect the human coastal communities that harvest, trade, and feed on these stocks. The increase of diseases in aquatic organisms that may affect humans is another problem associated with climate change. So is the need to change the boundaries of protected marine areas, since new habitats will require new preservation areas. At the top of the ocean's food chain, humans clearly are causing a decline in fish stocks by harvesting more fish than the fisheries can produce. To make matters worse, human-induced climate change and pollution are altering conditions for countless species, further undermining the ability of fish to reproduce in adequate numbers.

ESTUARIES AND COASTAL REGIONS

The shoreline—the boundary between land and ocean—is where the impacts of sea level, ocean circulation, and chemistry changes are the most conspicuous. Home to 50 percent of the world's population, coastal regions include beaches and headlands, estuaries and river deltas, wetlands and salt marshes. Our oceans' coastlines

Miami Beach is a typical example of overdevelopment right on the coast. This kind of development hampers the ability of the shoreline to adapt to change, and increases the potential for damages from sea level rise and storm surges.
© PETER ESSICK

support complex ecosystems, with a large diversity of plants, animals, birds, and insects, and provide nutrition and recreation to millions of people.

Many of the world's largest cities are built near the coast. Sea level rise, with increases in storm surge damage, endanger these metropolitan ereas. Tropical storm activity has increased in the Atlantic Ocean in the latter half of the twentieth century (although not in the Pacific; see Chapter 4 for details). In the North Atlantic, higher wave heights have been linked to more intense westerly winds in recent decades (Chapter 1). These factors increase the rate of coastal erosion, which washes away the shoreline.

Coastal erosion is already a widespread problem and has significant impacts on natural shorelines and on coastal development and infrastructure. Along the Pacific Coast of the United States, cycles of beach and cliff erosion have been linked to El Niño events that raise the average sea level and alter the path of storms in the region. For example, erosion damage during the 1982–1983 and the 1997–1998 El Niño events was quite widespread along the Pacific coastline.

The Atlantic Ocean and the Gulf of Mexico shorelines, as well as those of Southeast Asia, are especially vulnerable. Most erosion events on these coasts are the result of storms, and the coastal slopes are so gentle that a small rise in sea level produces a large inland retreat of the shoreline. Receding seashores threaten coastal development, transportation infrastructure, tourism, freshwater aquifers, and fisheries (which are already stressed by human activities). On the other hand, uncontrolled coastal development itself further increases storm damages by destroying sand dunes and beach vegetation that naturally protect against storm surges.

Subsidence, in which the land locally falls relative to sea level because of changes in the Earth's crust or changes in land or ocean deposition, may enhance coastal flooding even further. Subsidence also occurs due to human-related changes, such as dams and levees that alter water flow and reduce the sediment supply to the coastal region, and human activities such as groundwater extraction. The Mississippi Delta and Venice are two examples of places where subsidence is exacerbating the coastline's vulnerability to rising global sea level.

Estuaries are where rivers flow out into the sea, and of the thirty-two largest cities in the world, twenty-two are located on estuaries, including London, New York, Shanghai, and Buenos Aires. Estuaries are extremely productive ecosystems that are affected in numerous ways by climate change and other human activities. Increases in midlatitude rainfall enhance the volume of river water flowing to the sea, which further stratifies the waters where the fresh river water and the underlying salty ocean water mix. The river water may carry greater nutrient outflow from agricultural activities (particularly nitrates and phosphorous). These chemicals feed

ocean-dwelling algae and lead to increases in large algae concentrations (blooms). As the algal blooms decompose, the surface waters are depleted of oxygen, creating temporary dead zones where no other sea creatures can live.

New York City is located on the estuary of the Hudson River and has substantial amounts of reclaimed land that lies near sea level.
© JOSHUA WOLFE

Coastal wetlands (marshes and mangroves) are very important to the productivity of fisheries. Wetlands provide crucial nurseries and habitats for many commercially important fish and shellfish populations. Dramatic losses of coastal wetlands have already occurred along the Gulf of Mexico coast. Louisiana alone has been losing land at rates of between 60 and 100 square kilometers (24 and 40 square miles) per year during the last forty years, accounting for as much as 80 percent of the total U.S. coastal wetland loss.

In general, coastal wetlands survive if the rate of soil buildup equals the rate of relative sea level rise, or if the wetland is able to migrate inland. However, if sea level rise is rapid, or if wetland migration is blocked by bluffs, coastal development, or shoreline protective structures (such as dikes, seawalls, and jetties), the wetland

The outflow region of the Everglades in Florida is a typical example of a coastal wetland made up of mangrove swamps. It provides protection for the hinterland and a unique environment for the species that live there.
© PETER ESSICK

will be excessively inundated and eventually lost. The projected increase in the current rate of sea level rise will very likely magnify coastal wetland losses around the world, although the extent of the impacts will vary among regions.

OCEAN SURPRISES

It has been said that we know more about the surface of Mars than we do about the depths of the oceans, and that we make new discoveries and reach new understanding about the depths at a prodigious rate. This implies that the potential for us to be surprised by some of the consequences of climate change is probably quite high.

Perhaps one of the most extraordinary discoveries of the past two decades is the enormous amount of methane (CH_4) deposits that are frozen into a type of ice called methane clathrates. A clathrate is made up of a cage of molecules that traps other molecules inside. In this case, the cage is made up of water (and so the ice is more specifically known as a hydrate) and the molecule inside is CH_4. Hydrate deposits are found within ocean sediment along continental shelf regions of the world, but mostly in the Arctic Ocean and the Gulf of Mexico, where the pressure is high enough and the temperatures cold enough (less than a few degrees Celsius) for the hydrate to be stable. The amount of carbon held in these deposits is estimated to be comparable to all known reserves of traditional fossil fuel deposits. The key fact concerning hydrates is that they can be destabilized as temperatures increase. A warming ocean therefore may result in the melting of hydrates, leading to the escape of CH_4 into the atmosphere and escalating greenhouse gas levels. Deposits in the Arctic Ocean are possibly most susceptible because the Arctic region is warming faster than the rest of the world (see Chapter 2).

Another surprise often discussed is the possibility of dramatic changes in the thermohaline circulation. Ongoing changes have already been reported, but recent observations have revealed a great deal of previously undetected variability in this circulation. This means that some recent and well-publicized claims that an ongoing slowdown is taking place in the North Atlantic are not as significant as may have been thought originally because there is more noise in the system than signal. Currently, no good evidence exists that any changes are underway, but that does not rule out long-term changes occurring in the future. Large-scale monitoring programs have been initiated over the last few years that should provide a better basis for judging possible trends in the decades to come.

The uncertainties in these findings underline just how little of the world ocean really has been understood. Despite the heroic journeys of our plastic ducks, much remains to be explored.

Kim Cobb

LETTER FROM PALMYRA

My research has allowed me to dally on the most pristine beaches on the planet. It also has caused me to risk my life on several occasions. Such highs and lows characterize my brand of "Xtreme Science," a phrase my mother coined after hearing some of my stories from the field, even though the bulk of my research is conducted from the safe yet relatively boring confines of a university.

As a paleoclimatologist, I use the skeletons of corals from tropical Pacific reefs to reconstruct ocean temperatures for the last several millennia. Such information helps to define a baseline for natural climate variability, against which anthropogenic climate change can be measured. The coral-based climate data help to fill a large gap in the instrumental record of climate that exists in the heart of the Pacific Ocean. Climate in this region is heavily influenced by the El Niño/Southern Oscillation, a quasi-periodic cycle in tropical Pacific ocean temperatures and winds that occurs every two to seven years. The warm phase, referred to as El Niño, brings torrential rains to the west coast of the Americas and severe drought to parts of Australia, Africa, and Southeast Asia. Its cool sister, La Niña, has largely opposite consequences for global climate. The hope is that by looking at a detailed, long history of El Niño, as provided by corals, climate scientists are better equipped to understand the behavior of this relatively unpredictable yet critically important component of the Earth's climate system.

My fieldwork involves drilling cores from both living and dead coral heads located on remote atolls. Palmyra, Fanning, and Christmas are islands that lie along a north-south transect 2,500 kilometers due south of Hawaii. Palmyra and Christmas are accessible by plane from Hawaii, but the small twin-engine planes that service these islands cannot accommodate our 300 kilograms of gear, so we usually go by boat. We have cruised on a wide variety of vessels, including a mega-yacht donated by a prince of Saudi Arabia, a Norwegian Cruise Lines cruise ship, and a spunky yet dilapidated oceanographic research vessel. While not all vessels can top the amenities offered by the mega-yacht (including a private bathroom complete with bidet, a BBQ soirée with ten former Miss Tahiti winners, and a score of crisply uniformed sailors catering to my every whim), they all offered fantastic ocean views and, much more important, access to the tropical paradise that is my research site.

As pleasant as my fieldwork photos look (I have a long list of students, relatives, and friends eager to volunteer for my next expedition), the field trips are physically and psychologically taxing. For one, my plans for each field mission are always overambitious, such that personal downtime is sacrificed in the pursuit of constant dusk-to-dawn workdays. The work is grueling, involving the transport of hundreds of kilos of unwieldy gear several kilometers over uneven, heavily vegetated terrain. Add in a hot tropical sun, a 10-horsepower hydraulic drill, and the occasional scuba dive, and conditions can quickly become life-threatening. Indeed, one of my field partners was admitted to the hospital for four days upon his return to Hawaii because of a severe sunburn that had become infected as it festered in hot seawater for ten days. Heatstroke resulted in the emergency evacuation of two cruise companions. A light graze against a coral developed into a serious infection requiring powerful antibiotics. When you are a five-hour's flight away from a hospital, safety becomes the top priority.

Drilling a fifty-year-old coral head. © ZAFER KIZILKAYA

Finding fossil corals washed onto the shore at Palmyra. © JORDAN WATSON

Of course, nature can interfere with the best-laid plans, as I learned while collecting hourly seawater samples from a small powerboat in the middle of the large Christmas Island lagoon one night. My assistant and I were to spend seventy-two consecutive hours sampling seawater for a geochemistry study, operating off of a 10-meter boat anchored in 3 meters of water. The large yacht would moor just outside the lagoon, within sight, and we traded radio communiques on the hour all day and all night. Things were going splendidly (if a little sleepily) until the last night, when a serious squall developed. The wind picked up quickly, the rain and spray was so intense that we lost sight of the yacht's lights, and the radio was nothing but static. We huddled down below until it occurred to me that, for all we knew, we could have been blown into the open ocean by now. I decided to swim down the anchor line and make sure it was well secured. It wasn't, and we were being slowly pushed out to sea. In my panic, I tried to start the engine so that we could motor against the wind, but it wouldn't start (I would learn later that I likely flooded the engine in my desperation). Needless to say, I had to reset the anchor half a dozen times over the next hour as the storm was unrelenting. When day broke, we were astonished to see that we had been dragged almost half a kilometer and were less than 100 meters from the edge of the lagoon.

Nature would dole out several more nerve-wracking reminders of its supremacy and indifference over the years; I learn a new lesson on every trip.

Once back at the laboratory, we analyze the coral for geochemical variations that are linked to ocean temperature variations. The 3-meter-long cores from living corals extend back into the late nineteenth century, and are dated by counting back the annual growth layers visible in coral X rays. The fossil coral cores can date back six thousand years. So far, our data confirm that climate change of the past several decades dwarfs natural variations of the last millennium, in both the rate and magnitude of change. But the corals can uncover new mysteries about the tropical Pacific climate system, like a thirty-year period devoid of El Niño or La Niña events. This gap seems unnatural to a climate scientist, but it has occurred several times in the recent past, for reasons as yet unknown.

Our coral results suggest that climate change has accelerated in the tropical Pacific since the 1970s, which has immediate consequences for our understanding of the regional impacts of continued global warming. A warmer central Pacific might lead to warmer and wetter than average conditions along the west coast of the Americas, potentially influencing fisheries and water resource management. However, the local mark of rising tropical Pacific temperatures can be clearly observed by a short snorkel on the reefs of Palmyra, Fanning, or Christmas. These reef ecosystems have undergone a rapid decline over the last decade. When I first snorkeled on these reefs, on a cruise in 1997, I was struck by the diversity of coral and fish, and remember drooling over many coral heads several meters tall. On my last trip, in 2005, I was struck by the sparse coral coverage and the complete lack of coral heads larger than 1 meter in size. They had presumably been bleached and then eroded away by a combination of disease, storms, and grazing. Palmyra has been uninhabited since World War II, when it was used as a stopping-off point by the U.S. Navy, so the coral decline cannot be ascribed to the sparse populations that exist on Fanning and Christmas islands. Indeed, the inhabitants of these islands may be some of the first to lose their low-lying homes to sea level rise, just as their reefs succumb to the pressures of increasing ocean temperatures and accompanying disease.

For my part, I am personally devastated by the fact that the ocean may quietly and steadily drown Palmyra over the next century as sea level rises if we continue with business as usual. I can't help but think that a visit to Palmyra would inspire leaders to act on reducing carbon dioxide emissions, just as my first trip inspired me to become a paleoclimatologist.

CHAPTER 4

GOING TO EXTREMES

Adam Sobel

Blow, winds, and crack your cheeks! Rage! Blow! You cataracts and hurricanoes, spout till you have drenched the steeples, drowned the cocks!

—Shakespeare, *King Lear*

If you saw a heat wave, would you wave back?

—Stephen Wright

Climate change is the overall trend of the entire Earth's climate toward a different average state. It is most easily understood and predicted when framed in terms of averages over the entire Earth and over decades or centuries. These averages are changing, gradually by the standards of a human lifetime but very rapidly on a geological timescale. As individuals, communities, and nations, our experience of climate is much more local and immediate. We experience the climate through the weather. The most important weather events, for most of us, are the extremes. Our lives can be strongly affected by big storms, floods, and droughts.

Weather is capricious and random. On a day-to-day basis, its behavior doesn't have to be dictated by the global forces that are slowly forcing the climate to change. If climate is your average bank balance over a period of years, weather is the day-to-day fluctuation in your balance as your paychecks come in and your expenses go out. Let's assume you are fortunate enough to live in the developed world and have a high enough income to be able to buy books like this one. If someone put an extra dollar in your account every day, it might add up eventually, but you wouldn't notice much difference in those daily fluctuations. Assuming nothing else changes, those ups and downs would still be governed by whatever your income was before, minus your same daily needs and wants, because for most of us, these inflows and

Thunderstorm in Arizona. The maximum amount of water vapor in the air increases when the temperature rises. Intense downpours, for instance from large summer storms, can remove a large fraction of that water. Observations suggest that intense rainfall events are increasing in many parts of the world in line with increases in water vapor.
© GARY BRAASCH

outflows fluctuate by much more than a dollar a day. Similarly, greenhouse gases have the kind of gradual influence on climate that adds up over time. Although their influence on climate is inexorable, their influence on weather is not immediately detectable. Because of this, we can't generally say that any one extreme weather event was caused by climate change. Climate change will, however, have an impact on the statistics of weather. It will make some events a little more likely, and others a little less likely. Thus, we are sometimes able to say that the probability of certain kinds of extremes might increase, but since extremes are by definition rare events, detecting long-term changes in their likelihood will take time.

Our current understanding of extremes and their relation to climate change is very incomplete. It's still too early to use observations to say with certainty how extremes are changing. So we also use our understanding of the basic science, as well as computer models, to try to determine how we might expect extreme events to change as the Earth warms. What we can say at present consists of a combination of these expectations and an assessment of what has actually happened on the Earth so far. The results crucially depend on which sort of extreme we're discussing.

TROPICAL CYCLONES

Tropical cyclones are called hurricanes when they form in the Atlantic or the eastern Pacific oceans, typhoons when they form in the western Pacific, or simply cyclones when they form in the Southern Hemisphere or the Indian Ocean. But in all of those places they are fundamentally the same thing: intense large-scale vortices, swirling in the same direction as the Earth's rotation (counterclockwise in the Northern Hemisphere, clockwise in the Southern Hemisphere), that form over tropical oceans. They are warmer in the center than at the perimeter, and are accompanied by strong thunderstorms. Their energy comes from the warm ocean surface and the release of latent heat into already humid air as the water vapor condenses.

Tropical cyclones are among the most destructive natural disasters, particularly in terms of economic damage. Hurricane Katrina, which struck the Gulf Coast area near New Orleans in 2005, ranks as the most expensive natural disaster in the history of the United States. In many parts of the world, loss of life due to tropical cyclones is now relatively modest compared with occurrences in the past, largely because of improved forecasts. At the same time, the disruption to victims' lives due to the loss of homes, workplaces, and infrastructure can be enormous, as can the economic costs of these losses.

Is the number of tropical cyclones that occur every year increasing? Are the storms becoming more intense? Will places that have never seen a tropical cyclone

Hurricane Jeanne in the Atlantic near Florida in 2004. The spiraling bands of rain clouds and a well-defined eye are typical of these tropical storms. In this image, Jeanne is a Category 3 hurricane with sustained winds of over 160 kilometers per hour (100 miles per hour) and is about to make landfall on the Florida coast only 2 miles from where Hurricane Frances came ashore three weeks earlier.
NATIONAL OCEANIC AND ATMOSPHERIC ADMINISTRATION

in the past become threatened by them? While anthropogenic (human-induced) climate change itself is now an established fact, its influence on tropical cyclones is still subject to much uncertainty and debate.

Much of that debate revolves around the limitations of the historical data. The numbers and strengths of tropical cyclones naturally fluctuate over time, and we need a long record to determine whether all of those fluctuations add up to a long-term trend. Since the 1970s, the Earth has been observed by satellites, which make it relatively easy to see where all the tropical cyclones are at any given time. During this period, we have been able to record the number of storms each year quite accurately. However, although satellites can easily see clouds, water content, and rainfall, they have a much harder time judging intensity, or the overall strength of a storm. Storm intensity is measured by the maximum surface wind speed or the minimum sea level pressure. The intensity is best measured while flying an

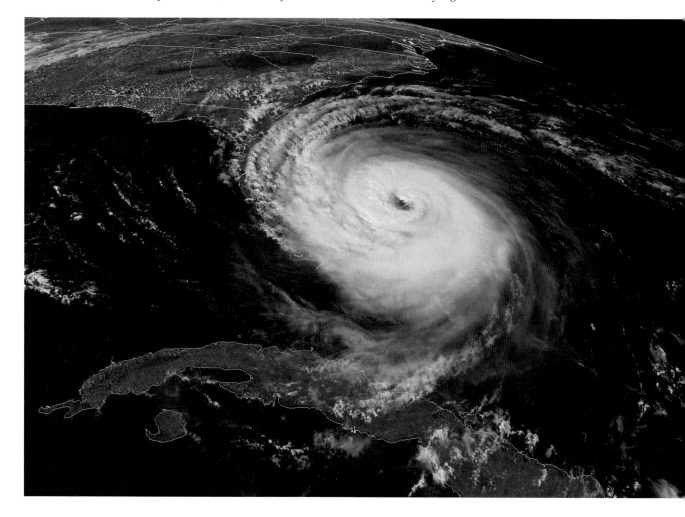

A sand flood in a beach home on the east coast of Florida after storm surges associated with hurricanes Jeanne and Frances. Jeanne followed Frances, making landfall within a few days of each other in September 2004. © GARY BRAASCH

airplane through a storm, which has been done routinely in the Atlantic for the last few decades, but has been done less systematically in other regions. Even in the Atlantic, aircraft reconnaissance hasn't always been frequent enough to give continuous data over the whole life of every storm.

Since the 1970s we have gathered good—but still not perfect—measurements of storm numbers and intensities over the Atlantic, and we know the numbers, but not necessarily the intensities, of storms elsewhere. Before this period, estimates of the number of storms and their intensity are less accurate, more so the further one goes back in time, and are almost nonexistent before about 1850. It's likely that some storms are missing from the record entirely, especially if they occurred over ocean regions without many inhabited islands or much ship traffic.

In the Atlantic, where the changes since 1970 are clearly detectable, we do have good enough data on both intensity and number to make some assessment of the variability. Since at least the early part of the twentieth century, and perhaps as far back as the mid-nineteenth century, a couple of peaks and valleys in hurricane activity have been noted, with each trend up or down lasting around a decade or two. The 1950s and 1960s were active, the 1970s and 1980s were relatively quiet, and the 1990s and 2000s have been active again.

These slow changes coincided with other changes in the climate system; for example, in the Sahel region of Africa, the 1970s and 1980s were a time of severe drought, following a wet period in the 1950s and 1960s, and the 1990s and 2000s have seen a partial recovery from the drought. Some argue that these variations—both in Atlantic hurricanes and Sahel rainfall—are part of a natural cycle, perhaps driven by slow fluctuations in the ocean circulation. By this reasoning, the strong upswing in Atlantic hurricanes since the 1970s is natural, and will be followed by a downswing eventually.

At the same time, both experience and an understanding of the physics involved tell us that the occurrence and intensity of tropical cyclones are related to sea surface temperature. For a given global climate, and all other things being equal, a warmer local sea surface corresponds to a greater likelihood of tropical cyclone formation. Once these storms start, the warmer it is, the more intense they can become. The increase in Atlantic hurricanes, for example, has occurred during a period of steadily increasing sea surface temperatures. Much of this warming is attributable to increasing greenhouse gases (see Chapter 6). There are also indications that the relatively cool Atlantic sea surface temperatures of the quiet hurricane period during the 1970s and 1980s may have been caused by anthropogenic aerosols (see Chapter 6). According to this line of argument, the fluctuations in both sea surface temperature and hurricane activity in the Atlantic during the twentieth century, including the increases since the 1970s, are not natural but "forced" by external factors acting on the climate, in this case aerosol and gas emissions from human activities.

If the increase in tropical cyclone activity in the Atlantic were related to forced increases in sea surface temperature, we might expect it to be accompanied by similar increases in tropical cyclone activity elsewhere in the world. Sea surface temperature has been increasing consistently over most of the tropical oceans, not just the Atlantic. Unfortunately, the data may not be adequate to determine whether other tropical cyclone regions are undergoing long-term increases in tropical cyclone activity. The most recent analyses of satellite data from the early 1980s to the present suggest that other regions have not seen the increases that the Atlantic has during this period. But we need longer, more consistent records to be able to discern the presence or absence of long-term trends more clearly. Nonetheless, it does seem at present that the clear, strong upward trend in tropical cyclone activity is mainly happening in the Atlantic. Though we don't completely understand it, this trend would bode ill if it were to continue.

What do we anticipate in the future? The degree of global warming we have experienced so far, though significant, is small compared to what we can expect. Even if the modest warming up to the present hasn't clearly led to global increases in tropical cyclone activity, should we expect that eventually it will? Since tropical cyclones prefer warm water, and global warming is warming the oceans all over the Earth, should we expect that there will be more tropical cyclones, and more intense ones, in the future? We aren't able to fully answer these questions yet.

To understand how the numbers of tropical cyclones may change in the future, we first need to better understand why the current numbers are what they are. We do understand broadly why tropical cyclones form over warm oceans and not over cold ones, but we don't understand the details well. About ninety tropical cyclones per year form in total over the entire Earth. This number doesn't seem to be changing significantly from one decade to the next. Why is it ninety, and not twice as many, or half? We don't know. The same thing applies, to some degree, to individual regions. The North Atlantic, for example, whose hurricanes threaten the United States, averages around ten tropical cyclones per year (with considerable fluctuations; a quiet year might have only five, and 2005 had twenty-eight), whereas the western North Pacific averages closer to thirty. We do understand broadly why the Pacific has more than the Atlantic does: a much larger chunk of ocean is warm enough for tropical cyclone formation in the Pacific than in the Atlantic. But why the averages are ten and thirty, and not five and fifty, or fifteen and twenty, we don't know. Without understanding the reason for the current numbers, it is hard to say how these numbers will change as the climate changes.

Besides sea surface temperature, other factors known to influence tropical cyclones, including vertical wind shear (the difference in speed or direction of the winds blowing at different heights), may be important. But we don't have complete confidence in our ability to predict how all of these factors will interact. Many current climate models do predict, however, that vertical wind shear will increase over the North Atlantic as a result of global warming. More shear would tend to reduce hurricane activity, counteracting to some degree the increases we expect as sea surface temperatures continue to increase.

Evaporation of water from the sea surface feeds a storm's rainfall. That evaporation depends not only on the water temperature itself, but on the humidity of the air just above it. For a given sea surface temperature and surface wind speed, the higher the near-surface humidity, the weaker the evaporation. When the water vapor in any part of the atmosphere reaches a certain saturation point, the vapor condenses (to liquid droplets, or ice if cold enough). The higher the temperature, the higher the specific humidity can be before that saturation point is reached. In a warmer climate, we expect the *specific* humidity of near-surface tropical air to increase as its saturation point increases, even as the *relative* humidity stays about the same (see On Commonly Used Terms for more details). This bit of physics means that the sea surface will have to be a little warmer than it is now for a tropical cyclone to form in the air column above it. We can't say, then, that tropical cyclones will form in regions that are now too cold for them to do so, just because they'll be warmer. That doesn't follow from what we know, although we can't authoritatively refute it either.

We know a little more about the intensity of tropical cyclones than we do about their number. Given some environmental conditions—the sea surface temperature and the temperature, humidity, and wind in the atmosphere, all of which we can predict to some degree with climate models—we have enough understanding of the physics of tropical cyclones to predict how strong a tropical cyclone in a given environment can possibly get. Most storms don't actually become as strong as this prediction allows, but this potential intensity—the maximum possible intensity of a storm—is the worst-case limit. The data from past storms indicate that if potential intensity increases by 10 percent, the intensity of the average tropical cyclone will also get about 10 percent stronger, even though it still doesn't reach its potential. It is very probable that the potential intensity will in fact increase in many places as the climate warms due mainly to the increasing ocean temperature. Thus, even though we don't know whether there will be more tropical cyclones or fewer, we do expect that, on average, the ones that do form will be more intense.

In the end, whether tropical cyclones become more frequent or stronger in the future is not the most important issue in determining whether the damage done to people and infrastructure by tropical cyclones will increase. From the early twentieth century to present, the economic damage caused by hurricanes in the United States (the region in which the most careful studies have been done) has increased tremendously. This rise in economic losses is much more than can be explained by changes in the frequency or intensity of hurricanes themselves. The real culprit is the infrastructure development along coastlines.

We keep putting more infrastructure in harm's way along the coasts in hurricane-prone regions. These increases in coastal development are a much larger factor than any changes in hurricanes that can reasonably be expected in the future. Even in the absence of global warming, we will still be in ever-increasing danger from tropical cyclones, as long as current development trends continue. One way to certainly reduce the harm done to our cities and towns by tropical cyclones would be to move human activity farther from the beaches.

In regions that depend on subsistence farming, such as southern Ethiopia, food shortages can occur in drought-stricken years.
© PETER ESSICK

OPPOSITE: Satellite images of Lake Chad over time (1973, 1987, 1997, and 2001) illustrate just how much smaller this relatively shallow lake has become since the 1970s. It is currently less than 10 percent of its size in the 1960s. NASA/GODDARD SPACE FLIGHT CENTER SCIENTIFIC VISUALIZATION STUDIO

DROUGHTS

The Sahel is the part of western and central Africa that lies between the wet tropical jungles of the Guinea coast, to the south, and the bone-dry Sahara desert to the north. During most of the year the Sahel is dry. During the period from July to September, the rains move up from the Guinea coast into the Sahel, bringing what is known as the West African monsoon. The rain that falls during this time of year is critical to the livelihoods of the people in the countries of the Sahel, many of whom are subsistence farmers. There is little irrigation in these countries, so without rain, the crops do not grow.

In some years, the monsoon rains are weak or nonexistent, which leads to great hardship. The loss of crops leads to famines. The lack of water itself for basic sanitation and the reduced access to pure drinking water can lead to disease. During the 1970s and 1980s, the Sahel had a long string of drought years, with disastrous consequences. Since then, the rains have returned somewhat, though not to pre-drought levels.

A visible and obvious sign of this drought is what has happened to Lake Chad, a huge, marshy, freshwater lake in the eastern Sahel that borders the nations of Chad, Cameroon, Niger, and Nigeria. Millions of people in these countries get their drinking water from the lake, and many also depend on it for fishing and agriculture. The lake is very shallow—one to a few meters in most places—and small changes in depth can lead to large changes in lake area. After the wet period in the middle of the twentieth century, the lake covered an area of 26,000 square kilometers, or about 6.5 million acres. Since then Lake Chad has shrunk to less than one-tenth that size, leaving enormous areas dry that were underwater not so long ago. Although some of this drying may be due to increased diversion of irrigation water from the rivers that run into the lake, much of it is due to reduced rainfall.

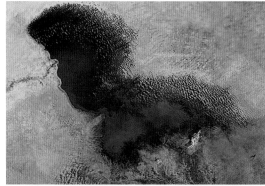
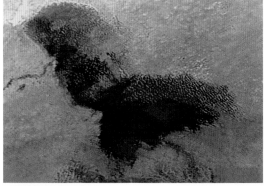
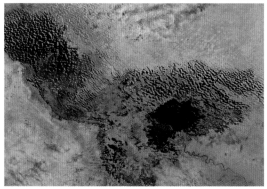

The Sahel drought of the 1970s and 1980s occurred simultaneously with other climate changes in nearby regions. During this period, compared with the 1950s and 1960s, the sea surface temperatures were warmer almost everywhere in the world except the North Atlantic. Both the global warming of sea surface temperature and the local cooling in the North Atlantic have been linked to anthropogenic emissions—of greenhouse gases in the former case and aerosols in the latter—though variability in ocean circulation is also potentially a factor.

Unlike tropical cyclones, large-scale droughts are phenomena that current climate models are, arguably, capable of simulating with some fidelity. When we give these models the sea surface temperature history of the twentieth century and simulate the atmosphere's response to those sea surface temperatures, we find most models respond by simulating a drought in the Sahel during the 1970s and 1980s that follows a wetter period, though in most cases the simulated drought is not as severe as the real one was. We can also run these models in a coupled mode, which simulates the behavior of the ocean as well as the atmosphere. Hence, sea surface temperature is part of the simulation rather than being required to match the observed history. To do so, we give the models our twentieth-century histories of greenhouse gas and aerosol emissions. When we run the models to simu-

Boat on the edge of Lake Chad. As the lake has shrunk in size, due mainly to the Sahel drought, weeds and grasses have grown up where the lake waters used to be, making access to the lake increasingly difficult for the 20 million local residents. © PETER ESSICK

late both ocean and atmospheric conditions, most of them again reproduce a Sahel drought (although again one that is too weak), as well as the sea surface temperature pattern that went with it.

These simulations, driven by histories of greenhouse gas and aerosol emissions, imply that the drought in the Sahel was in part caused by anthropogenic emissions. Of course, if those emissions were not present, variations in Sahel rainfall from year to year and decade to decade, and periods of drought, would still occur. Nonetheless, we think the Sahel drought may have been an early example of large-scale human suffering induced, or at least exacerbated, by anthropogenic climate change.

The Sahel is not the only part of the world prone to drought. Most of our planet's deserts are found in the subtropics, the latitudes not too close to the equator, but not too far either. Many parts of the subtropics are either chronically dry or fluctuate between dry and wet periods.

The American Southwest is dry compared to the eastern United States, but just how dry varies over time. By looking at the rings of ancient trees in the Southwest, scientists have been able to reconstruct the climate history of this region over the last thousand years. This history shows relatively wet periods, and periods of drought that were more severe and long-lasting than any that have occurred since Europeans came to the region. The ancient tree trunks that have been found upright at the bottom of modern-day riverbeds show that these beds must have been dry land when the trees grew—spectacular evidence of these early megadroughts.

The most severe drought in the recent history of the United States was the Dust Bowl of the 1930s. During a relatively wet period in the American West in the late 1800s and early 1900s, the U.S. government encouraged people to move west. However, over the next few decades, the settlers were faced with a climate changing to much drier conditions. As with the Sahel drought, climate model simulations have been able to reproduce the Dust Bowl when given the history of sea surface temperature. In this case, though, the drought was clearly not anthropogenic, but natural, since it occurred before human emissions of greenhouse gases had yet had a significant influence on the climate. However, the effects of the drought were greatly exacerbated by poor farming practices.

The necessary ingredient instead appears to be the natural occurrence of La Niña, a cooling of the eastern equatorial Pacific Ocean that is the opposite of El Niño (discussed in Chapter 3). The Dust Bowl years were a time of mild but persistent La Niña conditions, and the presence or absence of the La Niña event in climate models makes the difference between the presence and absence of the drought. The same kind of experiments link La Niña to the other known American West droughts of the last 150 years.

In recent years, the American West has again been subject to drought. Water management infrastructure in the Southwest, planned during wetter times, is stressed by low flows in major rivers and low levels in both human-made and natural lakes. Unlike previous droughts, the current one does not seem to have been caused by La Niña. Although there have been a couple of La Niña events in recent years, they haven't been persistent enough to account for the drought. On the other hand, long-term projections of climate change indicate a shift to drier conditions in a number of subtropical regions, including the southwestern United States, the Mediterranean, and Australia (which is currently also in a long-term, severe drought), due to anthropogenic climate change. The mechanism of these droughts is different from that of the natural, La Niña—induced ones. Instead, it is

Lake Powell is an artificial reservoir behind the Glen Canyon Dam on the Colorado River. After seven years of drought in the southwestern states, it currently holds only 50 percent of the water it did in 1999. The rock is visibly marked by a "bathtub ring," a reminder of wetter times. © PETER ESSICK

related to a change in the atmospheric circulation in which the storm tracks (the paths that midlatitude storms follow) move a little farther from the equator and closer to the poles as the climate warms. As the storm tracks move poleward, the subtropical dry zones left behind get larger and drier. Exactly why this change should occur in a warmer climate is not entirely understood, but it is a consistent feature of projections made with many different climate models. Although we are still in the early stages of global warming, and again it is difficult to attribute the present droughts entirely to this warming, there are some strong indications that these droughts are indeed not just temporary, but rather may be shifts to a permanently drier state.

FLOODS

The opposite of drought is flood, but the two are not quite the mirror image of one another. Droughts develop slowly, over long periods of time, as rain that is expected to fall fails to do so, day after day, month after month, year after year. Floods often develop very quickly, in response to extremely intense rainfall that may occur over just a few days, or perhaps a few intense storms over a single season. If the same amount of rain falls in many smaller storms, as opposed to one or a few very large ones, the likelihood of flooding in many places is generally much smaller. Floods are true extremes, related to the heaviest rain events, rather than just the cumulative average of all of them.

If climate change causes an increase in floods in some places, we might not see it clearly for a long time, because the increase will really be just a small shift in the odds of what is already, in any given year, an unlikely event. To reliably detect such a shift, we would need to wait a long time, long enough for a large number of such events to occur so that we could get an accurate estimate of their frequency. Because of this purely statistical problem, we don't expect to be able to tell yet whether major floods have become more common as the Earth has started to warm. Some reports have indicated that there have been increases in more common floods—those associated with rain events that are big enough to be noteworthy, if not catastrophic. Such increases have been found to be occurring in many places, consistent with what we expect to happen. In regions that are currently flood-prone, we expect the frequency of high-intensity floods to increase in a warmer world. This expectation is based on not only climate model projections, but also basic physics.

As mentioned earlier, the higher the temperature, the higher the specific humidity (the actual amount of water vapor in a given amount of air) can be before the air becomes saturated and condensation occurs. Therefore, when condensation does occur at higher temperatures, the resulting precipitation can carry more moisture and have a greater intensity.

In global warming projections from virtually all modern climate models, the gross relative humidity patterns in the atmosphere stay nearly constant as the climate warms. If the planet warms by 2°C or 3°C (4°F to 6°F), we expect specific humidity near the surface to increase by 15 to 20 percent; however, the total amount of rain falling isn't expected to increase that much. In the global yearly average, rainfall totals must match closely the amount of water evaporated from the surface of the Earth. Evaporation in turn is controlled by other processes, including the amount of infrared and solar radiation reaching the surface, which aren't expected to change as fast as water vapor does (7 percent per degree Celsius of warming). We

A man searches for food inside a flooded house in Doncaster, England, after the unprecedented extreme rain events of June 2007. Rain has been falling in intense bursts in Europe more frequently in recent decades. © ASHLEY COOPER/ GLOBAL WARMING IMAGES

do expect global rainfall to increase as the climate warms, but perhaps only by 1 to 3 percent per degree Celsius of warming.

The global mean rainfall is not an accurate indicator of local trends in rainfall. The global mean is an average over a huge number of individual rain events, which themselves aren't directly controlled by the need to balance the budgets of water and energy for the whole globe. The upper limit on an individual rain event is simply set by the amount of water vapor in the air, meaning that the amount of rainfall in such intense storms could increase as fast as water vapor does, that is, at the rate of 7 percent per degree Celsius of warming. This is an upper limit that doesn't account for the actual dynamics of storms—it's possible that the rate of increase will be less. The minimum rate of increase is presumably that of global mean precipitation (1 to 3 percent per degree Celsius of warming).

Analyses have shown that the amount of rain falling in the most intense downpours has increased faster than mean rainfall amounts over the last thirty years in the United States, Europe, and India, and in some places, heavy rainfalls increased even when total rainfall was static or falling. Rainfall statistics are always noisier than those for temperature, and results are not available for many tropical and sub-tropical areas or over the oceans, so we can't use observations to confirm that these results are representative of the whole globe. Nonetheless, the simple arguments above suggest that further significant increases in rainfall will occur with a few degrees of warming. Such an increase would be particularly unfortunate for already flood-prone areas; it would be even worse for low-lying ones such as Bangladesh and the Netherlands, which simultaneously have to cope with rising sea levels.

HEAT WAVES AND COLD SNAPS

Understanding how the frequency and intensity of heat waves are changing as a result of global warming is just about as simple as you'd think it would be. As the planet gets warmer, there will be more heat waves, and worse ones.

Again, although individual events cannot be directly ascribed to long-term trends, recent years have seen a number of severe heat waves around the world. The devastating European heat wave of 2003 was a truly extreme weather event by the standards of the twentieth century, and the death toll has been estimated at over 35,000 people. Several heat waves in the last decade or two have had death tolls in the hundreds, such as the Chicago heat wave of 1995 and the heat waves of 2006 in both Europe and North America. The conditions experienced during these events may come to be the norm by the end of the twenty-first century. For people who don't live in regions that are drought-prone or on the coast, the most direct consequence of global warming may be simply the warming itself.

The flip side of the increase in heat waves, and another very straightforward consequence of global warming, is a decrease in extreme cold snaps. Such cold events have become less frequent in recent decades than they were previously, and this trend is very likely to continue. This change will be welcome in some places, because it will mean reduced heating costs, for instance. It may not be welcome in others, as some species of plants and animals need cold spells—for example, apple and sugar maple trees in the northeastern United States and Canada. Some species that are viewed as undesirable pests are kept out of high-latitude ecosystems by their lack of tolerance to the cold; these species, such as the pine bark beetle, now flourish in Alaska and British Columbia as very cold days and nights become less common.

The science of extreme events is still relatively primitive. Observational data is often lacking or of poor quality over the necessary time period, and so it is hard to find good tests to determine how accurate climate models are. Unfortunately, extreme

weather is the lens through which many people see climate change, since these are the events that get media coverage and have the most dramatic potential. However, not all extreme events are alike, and although uncertainties still exist, scientists have expectations that some extremes will become more likely, or more extreme, in the future. Some of these changes—particularly the increase in droughts, heat waves, and intense rainfall events—may already have been detected in observations. For some others, such as changes in hurricane number and frequency, the observational record is still murky, and only time will tell whether the expectations based on current understanding will pan out.

During the summer of 2007, two sets of forest fires raged across Greece. The first photograph is from Poros, where fires broke out during the period of highest temperatures (above 43°C, or 111°F) of the June heat wave. A second wave of fires occurred in August, as seen in the satellite image. A combination of scant precipitation and multiple heat waves left Greece particularly vulnerable. By the end of the summer, the tally was 120 major fires and 469,000 acres of forest land burned. ABOVE: © JOSHUA WOLFE; OPPOSITE: NASA

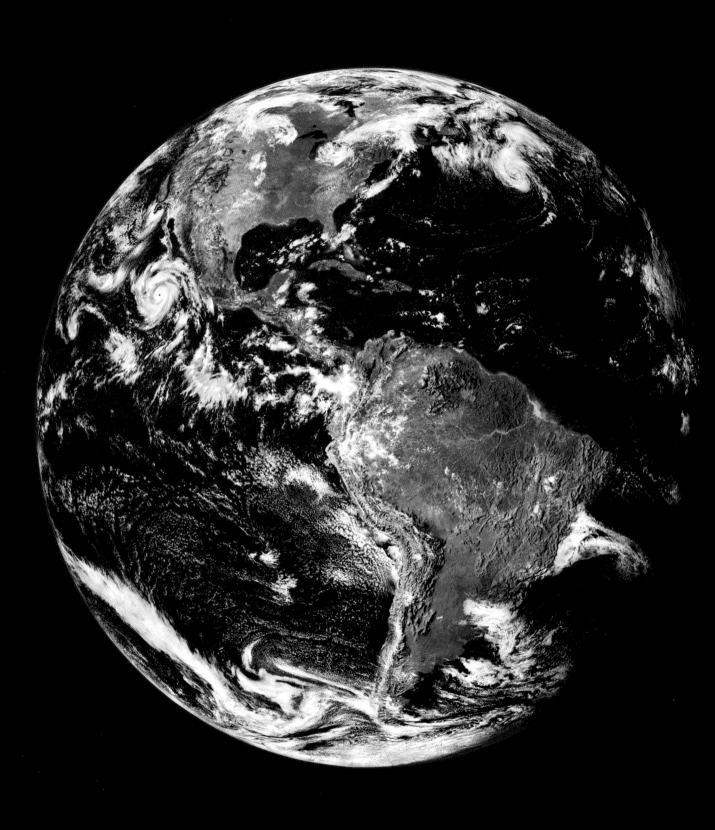

CHAPTER 5

THE LIFE OF THE PARTY

Shahid Naeem

Life is the fire that burns and the sun that gives light. Life is the wind and the rain and the thunder in the sky. Life is matter and is earth, what is and what is not.

—Seneca

The system of life on this planet is so astoundingly complex that it was a long time before man even realised that it was a system at all and that it wasn't something that was just there.

—Douglas Adams

Viewed from orbit, Earth is a spinning blue planet with three-quarters of its surface covered by water. Every year, a thin veneer of greenish brown land expands in one hemisphere as Earth's icy cap retreats for a few months. Six months later the same pattern is observed in the opposite hemisphere. The light we see with our limited vision, however, is only a very small portion of the full spectrum, ranging from X rays to radio waves and passing through the ultraviolet, visible, and infrared bands on the way. The spectrum is usually divided up into wavelengths measured in nanometers. X rays have extremely short wavelengths, radio waves have very long ones, and visible light is somewhere in the middle. If we were to examine the spectra of Earthlight more closely, we would notice two things: emissions at 300 nanometers (in the ultraviolet) are very weak, and reflectance of sunlight at about 450 and 700 nanometers would be muted compared to nearby wavelengths. The first case is associated with ozone in the stratosphere, which absorbs almost all the solar ultraviolet radiation, and the second is associated with the absence of red and blue light. These features are surprising. Atmospheric ozone is very reactive and can only be maintained with a persistent source of oxygen. And since the Sun that

Planet Earth. NASA IMAGE CREATED BY RETO STOCKLI WITH THE HELP OF ALAN NELSON, UNDER THE LEADERSHIP OF FRITZ HASLER

provides the light has plenty of reds and blues in its spectrum (as seen in a rainbow for instance) something must be preferentially absorbing these wavelengths.

What our observer cannot see from space is that trillions of individual organisms inhabit the Earth's surface. Collectively, the zone on the Earth's surface in which life is found is our planet's biosphere. These organisms are suffused throughout Earth's air (our atmosphere), Earth's oceans and seas (our hydrosphere), and Earth's rocks, sediments, and soil (our lithosphere). The inhabitants of the biosphere include the bacteria and fungi in the soil, the majestic canopy trees of the rainforest, all insects everywhere, the plankton swimming the ocean, and, of course, vertebrates such as birds, lizards, fish, and humans. The sum of the mass of all the plants, animals, and microorganisms within Earth's biosphere weighs a little under 1,000 billion tons. It should be no surprise that such an enormous living, growing, reproducing, breathing, rotting, and waste-producing mass hugely influences Earth's environment. Biological processes move billions of tons of elements, compounds, and materials among Earth's spheres every year. Plants and oceanic algae, for example, are responsible for the spectral anomalies mentioned above; they keep atmospheric oxygen levels high and carbon dioxide levels low by using the electromagnetic energy in visible light to pull the carbon out of the air. They do this using chlorophyll—an incredible protein that can collect energy across the visible spectrum—particularly in the red and blue bands, leaving the green wavelengths to be reflected back to our eyes and defining our everyday impressions of vegetation.

The biosphere has created this environment, and it is this environment that allows the biosphere to persist. The mass of living organisms on Earth has influenced the climate and is itself influenced by climate. The biosphere is undoubtedly a lead character in the climate change drama.

ORDER AND STRUCTURE IN THE BIOSPHERE

The sheer mass of the biosphere is impressive, but it is the diversity of life that truly astonishes. From a great distance, the Earth's biosphere seems a homogeneous mass covering Earth's terrestrial surfaces—deep green in wet tropical regions and light brown in more arid regions. But the closer we look, the more complexity we see. If we try to count how many different types of organisms are currently on Earth, we come up with an astounding number ranging between 10 and 30 million different species, depending on whose estimates we use. And this does not include all the varieties within a single species—all dogs, for instance, comprise just one species.

Because all groups of organisms change over time—sometimes evolving into entirely new organisms, sometimes disappearing entirely—scientists need a stable

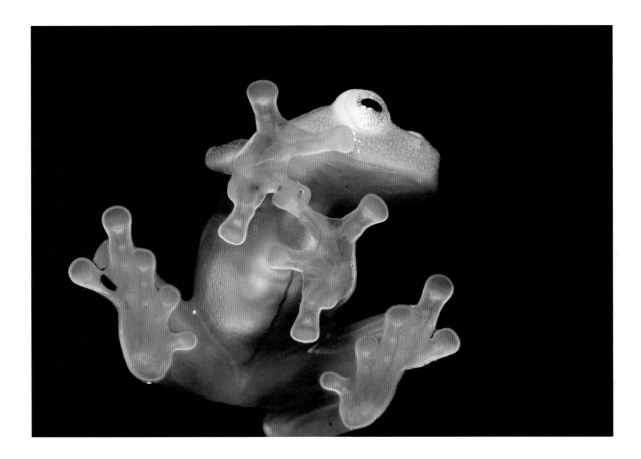

unit to measure diversity. Species, though opinions differ, are usually defined as an organism unit that breeds true for many generations. Indeed, the average lifetime from origin to extinction of a species, such as a specific kind of ant, mushroom, or oak tree, is about 4.5 million years. Given this relative stability, species are good units to count on when studying our biosphere's recent past, over decades or centuries, and thinking about its not too distant future. The Earth's massive biogeochemical engine, responsible for making the planet habitable, is made up today of 10 to 30 million working parts.

We seldom appreciate the significance of biodiversity (the diversity among living organisms) to our planet's biosphere and climate, because we are often mesmerized by life's extraordinary variety. At first glance, biodiversity seems an eclectic, enigmatic jumble of constantly evolving species with no particular rhyme or reason to its distribution or function on Earth. Protobacteria are a mere 0.5 microns long. Blue whales measure 25 meters (80 feet) in length. Redwood trees stand over 100 meters (330 feet) tall. Biodiversity conjures up a riot of form and function scattered about the Earth: lions, tigers, bears, fleas, mice, clams, fish, worms, mushrooms,

More than five thousand species of frog have been described. This one is an example of Fleischmann's frog, which lives in forests from Colombia to Mexico. It is one of many species that are threatened by habitat loss and climate change. © PETER ESSICK

Independent biologist Debra Hamilton surveys birds in the moist mid-level mountain forest of Monteverde, Costa Rica, using mist nets. The orange-billed nightingale thrush (in the net) is a species whose range has moved up in elevation since the 1970s to maintain its ideal environment. It will be tagged and released to allow for future monitoring of its range. © PETER ESSICK

and a mélange of other creatures big and small, wild and tame, colorful and dull. Some microorganisms live for only a few hours. Bristlecone pine trees may live for more than four thousand years. Sedentary invertebrates inhabit the dark abysses 5 kilometers below the ocean surface, and seabirds fly thousands of kilometers to circumnavigate the globe. Earth's diversity of life defies easy description.

Biologists spend most of their time cataloging, collecting, and displaying species in museums, gardens, zoos, private collections, and more recently, on an increasing number of Web sites. Cataloging life is an important endeavor, to be sure, but like any catalog of parts, it sheds little light on the configuration or arrangement and gives no particular insight into the structure of the system.

When we take a break from inventorying species to look at the spatial, temporal, and ecological properties of biodiversity, we discover something that is simply stunning. Life is surprisingly *ordered*. The riot of species and the seeming chaos of life on Earth belie the fact that biodiversity is a highly structured system. The first evidence of this is the banded pattern of biomes, the major climate-controlled ecosystem types. Around the globe, tropical rainforests occur in the equatorial belt, savannas (grasslands with scattered trees) lie to the north and south of rainforests, grasslands to the north and south of savannas, deserts to the north and south of grasslands, then grasslands again, followed by temperate forests, boreal forests (or taigas), and tundra up to the icy poles. Mountains or oceanic currents may make a region anomalously warm, cold, wet, or dry, so topography may muddy this well-ordered pattern. But if we know the pattern of temperature and precipitation in the region, we can tell what biome we are likely to find.

Herein lies the first key point: biodiversity has helped to create the Earth's anomalous climate, but climate governs the well-ordered patterns of distribution and abundance of plants, animals, and microorganisms on Earth. Climate and biodiversity are thus inextricably linked.

We use the term *ordered* in the sense that species are not randomly strewn about the world. Every species has a highly predictable, specific place and time of occurrence in the biosphere. Every species therefore has a well-ordered pattern of where it is distributed and how abundant it is in those locations. Wherever or whenever the physical and chemical conditions permit, species of some kind may be found almost everywhere on Earth. But no one species is found everywhere. Most species have very specific, narrow ranges of tolerance for physical and chemical conditions. The narrow tolerances for temperature or saltiness of seawater, for example, limit that species to particular locations at particular times. This envelope of conditions is what ecologists refer to as a *niche* and constitutes a sort of species address.

A second key point is that temperatures, moistures, and other environmental factors for which we find a well-adapted species that tolerates those conditions are very diverse. Earth is covered with living things—pick almost any location on Earth and you will find species that can survive there. For example, we had long thought no life would tolerate the tremendous heat, pressure, and toxic lack of oxygen found in the vents at the bottom of the ocean where sulfur-laden gases escape as the sea floor spreads. Yet we now know that even this seemingly inhospitable environment harbors life that thrives in this niche. Microorganisms can live in the most extreme environments. They live around hot, sulfurous hydrothermal vents beneath the sea, on the surface of sea ice in the Arctic, beneath the sterile sands of the Atacama Desert, and in the deepest trenches of the oceans. Microorganisms even can live in the air: tens of thousands of species of viruses, bacteria, plant and fungal spores, and tiny insects make up what is known as aeroplankton. Biodiversity is not just the remarkable variety of shapes, colors, and behaviors (intriguing though these are), it is the variety of ways species make their living on Earth.

The third point worth considering is that biodiversity is *structured*. Within any given location, each of the living organisms plays a specific role that helps support that ecosystem and without which the ecosystem would be weaker and more vulnerable to abrupt change. For example, oak trees bear acorns, squirrels eat the acorns or bury them, and the buried acorns grow into new oak trees. An ecosystem is not a random hodgepodge of creatures living in a chaotic free-for-all. Every species is linked to every other by its mutual use of the energy and nutrients flowing through or cycling in the habitat. Sometimes these linkages are rather direct, such as when a species "shares" energy by consuming another, either as a carnivore, herbivore, or parasite. While food chains and food webs are ubiquitous elements of structure in ecosystems, many other linkages are less direct, such as competition for resources, or its opposite, mutualism—when species work together to acquire the resources they need to grow. A well-known example of a mutualistic interaction is the way sedentary flowering plants provide nectar as food to mobile pollinators, which in turn carry pollen from one sedentary flower to the next, helping stationary plants procreate even though they are far apart. Even more spectacular is the symbiosis between coral polyps and zooxanthellae discussed in Chapter 3, in which the polyps provide shelter and nutrients for the zooxanthellae and the zooxanthellae provide energy through photosynthesis for the polyps.

Ultimately, all species of living organisms interact directly and indirectly because they share carbon, nitrogen, and other elements that are parts of global biogeochemical cycles. This massive connectivity of living things represents the structure of biodiversity and is critical to making the biogeochemical engine work.

CLIMATE CHANGE AND BIODIVERSITY

Keeping these three things in mind—that biodiversity both contributes to and is affected by climate, that species are spatially and temporally restricted to specific sets of environmental conditions, and that biodiversity is structured by interactions among species—what are the consequences of climate change to biodiversity? Fortunately, we can answer this question because people have been observing species for a long time.

For centuries people have studied where and when plants flower or birds migrate or breed or nest, a science known as phenology. Grape harvests in Europe have been tracked since at least the fourteenth century. Cherry blossom dates have been recorded in Japan for a thousand years. These observations note the shifts of such milestones through time and have allowed scientists to carefully and quantitatively examine the influence of recent climate change. What they have found is staggering. Almost anywhere anyone looks, on average, climate change in recent decades is changing the order and structure of life on Earth.

Should we be surprised? Paleobiologists, those who study the history of life and reconstruct what nature looked like in the past, have documented the extraordinary dynamics of biodiversity, which almost always involves responses to climate change. In colder epochs, glaciers expand and forests retreat; when the glaciers retreat again, forests expand. During the mid-Cretaceous period 90 million years ago, when the atmosphere was rich in carbon dioxide, the polar ice caps were absent, and the entire world was warm, crocodile-like creatures called champosaurs roamed what are now the waters of the Canadian Arctic, and tropical breadfruit trees were found in what is now Greenland. Scientists have discovered that, just six thousand years ago, large parts of the Sahara were well-watered grasslands with abundant wildlife.

Many past changes in climate were much larger than the changes we see today, and biodiversity responded to each of them, though responses sometimes took on the order of hundreds or thousands of years to develop. For example, in the last one hundred years, global mean temperature has changed only about 0.8°C (1.4°F). This rise seems inconsequential when one considers that Earth was 6°C to 8°C (11°F to 14°F) warmer back when champosaurs wandered around the Arctic. Surely biodiversity can cope with climate change? And surely current changes in climate have been too small to have yielded any significant response by biodiversity? In actuality, biodiversity appears to be having trouble in part because current climate change is occurring very quickly. When change is quick, even small differences can have dramatic impacts.

Climate change has two kinds of impacts on biodiversity. The first impact relates to which species are located where and when—the ordered nature of the biosphere. The second deals with the structure of the living elements of the biosphere—how the species in any given location interact with each other. Altering the order and structure of biodiversity by changes in climate should weigh heavily in our deliberation about our future.

CHANGES IN SPACE AND TIME

Changes in the way species are ordered are widespread, though it is more obvious for some species than others. Recall that species have specific conditions they can tolerate. When conditions change, species either move, adapt, or perish. The evidence we have today shows that all of these responses are occurring now or are likely to occur in the near future. One recent study that examined almost 1,600 species estimates that more than half have shifted either their geographic range or their phenology over the last 20 to 140 years.

The anole lizard of Selvatura Forest, Santa Elena, Costa Rica, is another species that is moving upslope as temperatures rise. © PETER ESSICK

In the Northern Hemisphere, midlatitude species have been moving north by an average of 6 kilometers (3.5 miles) every decade, while mountain species have been moving upslope 6 meters (20 feet) every decade. There is less area at the tops of hemispheres and the tops of mountains, so eventually, as warming continues, the preferred habitats of some species will completely disappear. Nearly two-thirds of North Sea fishes have shifted in mean latitude or depth over the last twenty-five years in response to local warming, with the majority showing northward shifts.

Vegetation is particularly sensitive to changing climate. When warming allows the invasion of trees, previously treeless tundra can convert to boreal forest. When either warming or deforestation damages rainforest areas, they can convert to grass-dominated savanna. Perhaps the best-known savannas are those of Africa, such as the Serengeti Plain. The llanos of the Orinoco basin of Venezuela and Colombia, the cerrado of Brazil, the pine forests of Belize and Honduras, and the eucalyptus of Australia are other examples. Fire, grazing by large mammals, flooding, and drought all are factors that can tip the balance between trees and grasses. For thousands of years during the last ice age, the Amazon basin was as much as 5°C (9°F) cooler in certain parts than it is today. The drier weather and more frequent fires of this long period favored savanna over rainforest, though there is debate about the exact extent. Ironically, over the long term, biodiversity in the Amazon increased because of these changes, as the evolution of new species appearing in the remaining rainforest refugia overcame the losses due to habitat change. Unfortunately, the current rate of change is so fast that evolution is not going to be able to catch up. Although projections vary, some climate change models suggest that the region will become warmer and drier. Fires set by humans and increased grazing by cattle could lead to a rapid reversion of the rainforest to savanna.

In the Southern Hemisphere, penguins that breed and depend on ice in the Antarctic, such as the Adélie penguin, have been declining in numbers since the 1970s near the Antarctic Peninsula. Warming trends in this region have led to less ice cover, reducing the abundance of krill, a vital staple in fish diets in the region. On the other hand, this alarming decline has been accompanied by increases in the chinstrap penguin and Gentoo penguin, which have expanded their range southward and arrived in the Antarctic Peninsula only in 1976 and 1994, respectively. In a further complication, analysis of fossil egg shells indicates that Adélie penguins only started depending on krill relatively recently. Prior to the nineteenth century they ate mainly fish. It was the almost complete annihilation of seal and whale populations (which normally eat krill) by human hunters that allowed the krill to expand and become easier prey for the penguins.

Adélie penguins on Humble Island off the Antarctic Peninsula. This region has seen some of the most dramatic warming in recent decades, and the impacts of changing weather, rookery snowfall, loss of sea ice, and reduced krill availability have been severe. © GARY BRAASCH

Changes in the phenology of species' biological activities are also evident. The growing season of many plants has increased in the northern reaches of the United States and Europe; several species of frogs have begun breeding earlier in New York and England; birds are laying eggs earlier in southern Arizona and the United Kingdom; migrant birds in the North Sea are migrating earlier, whereas in Spain they are arriving later; butterflies are appearing earlier in the United Kingdom, Spain, and California; lilacs and honeysuckle are blooming earlier in the western United States; and algal blooms are occurring earlier in lakes of the northwestern United States—and these examples are just in the places where we have looked and have the long-term data to assess the change.

Climate-induced changes in spatial distribution and timing of species together in a single community can have surprising, often worrisome, consequences. For example, when the migratory pied flycatcher arrives in the Netherlands to nest and raise its young, it now arrives too late for the peak emergence of the insects that serve as its food. In places where this mistiming is the worst, the local population of these birds has declined by 90 percent. This is an example of a tipping

CLIMATE CHANGE

point, which is a common characteristic of complex ecological systems—seemingly small changes in background conditions can have dramatic and disproportionate impacts.

A more complex example is the Mediterranean ecosystem of Europe. One study found that, over a fifty-year period, spring was coming earlier—leaf unfolding by sixteen days, flower opening by six days, and fruit appearance by nine days. In this same system, compared with 1974, butterflies appeared eleven days earlier, and migratory birds arrived fifteen days later. Given the differences in these temporal shifts, it is possible that flowering and the arrival of pollinators such as butterflies will no longer overlap for a sufficient period to allow pollination to take place.

Changes in the relative abundance of species can be observed as well. In the western Antarctic Peninsula, for example, on top of the shifting ranges of penguins described above, the number of salps (gelatinous invertebrates) found in the open ocean has increased, while the number of krill, killer whales, seals, and seabirds has declined. Because predatory species mostly favor krill over salps, a smaller krill population could lead to a decline in predatory fish, birds, and mammals, and also a decline in the microorganisms salps feed on as their numbers increase. Thus, everything from bacteria to whales may be affected.

Similarly, in the opposite hemisphere, the tiny species making up the marine plankton of the North Sea have showed shifts in the timing of peak abundances among different groups. One study found that over a forty five-year period (1958–

This Harlequin toad (*Atelopus varius*) in Costa Rica is one of dozens of this genus on the endangered species list (some of which are already thought to be extinct) due mainly to the spread of the chytrid fungus, possibly accelerated by climate change and harvesting for the pet trade. © GARY BRAASCH

2002) in which sea surface temperature increased 0.9°C (1.6°F), the dinoflagellate algae shifted by twenty-three days, the hatching of larvae of bottom-dwelling species that feed on algae shifted by twenty-seven days, and the predatory copepods and many other planktonic species shifted by ten days. Most of these species interact with one another in a complex food chain and are the source of food for larger organisms, so their mismatch in seasonality could spell trouble for the community.

Biodiversity's structure is also being disrupted, as is seen in the decline of harlequin toads (sometimes described as frogs) in the Central and South American tropics. Sixty-seven percent of these frogs have disappeared over the past twenty to thirty years, possibly because local climate change has created ideal conditions for the chytrid fungus infections that kill these and other tropical amphibians. Another example is that of the interaction between bark beetles and North American pine trees. Bark beetles are a species native to North America, but increases in temperature over the last fifty years have favored their growth and survival while drought has weakened the defenses of pine trees against the beetles and the pathogenic fungi they transmit. In some regions, such as parts of Colorado, up to 90 percent of the pine trees have been attacked and killed.

The most dramatic example of a direct disruption of a species interaction by climate change is the breakdown of the polyp-zooxanthellae mutualism in corals described in Chapter 3. More than a quarter of marine invertebrate and vertebrate species are associated with coral reefs, thus an enormous fraction of marine diversity can be impacted by the disruption of just this single element of biodiversity's structure.

The number of documented impacts of climate warming is increasing each year, but we can only document changes in the activities of plants and animals that we have been watching. In most cases, we are not able to document how sensitive biodiversity is to climate change because we just haven't been paying attention. Some of these changes are observable from space, where one can see the increased growing season in satellite images of the far Northern Hemisphere as it is staying greener for longer periods.

What we have seen accords with the *kinds* of changes biologists have predicted, but the degree to which they have occurred has been surprising. As climate change progresses, the number of documented alterations in biodiversity will increase enormously.

HUMANS AND CHANGING BIODIVERSITY

The likely loss of ice-dependent (and photogenic) species such as polar bears and Adélie penguins, the potentially adverse effects of declines in krill on whales, the loss of corals and coral reefs to bleaching and storms, and the loss of many other species that cannot move or adapt, or have no place to go when climate change occurs, may upset many people. The impacts of climate change, however, go beyond our cultural appreciation of biodiversity. Ecosystems provide specific benefits called ecosystem services. Recreational or inspirational values of biodiversity, such as whale watching, snorkeling in coral reefs, or even simply knowing that penguins exist, are some examples of cultural ecosystem services. Other ecosystem services, however, include supplying provisions that humans need, such as food, fuel, clean water, and fiber. Healthy ecosystems provide other benefits that are supporting services, such as producing oxygen and sequestering atmospheric carbon dioxide. Other ecosystem roles include regulating services, such as coral reefs stabilizing coastlines or forests contributing to holding greenhouse gases in check. These latter, more utilitarian services, are ultimately more important than the cultural benefits we enjoy. For the Inuit who annually harvest thousands of ringed seals and hunt polar bear for meat, clothing, and tool- and boat-making material, the decline of seals and bears due to the retreat of the northern ice is an existential threat.

Another example is the shift in seasons for maple trees in North America. In the United States, the northward shift in warmer climates may lead to greater pro-

ductivity in some species. But for maple syrup harvests, mild dry winters, the shortening of the transition from winter to spring, and in particular, the reduction in freeze-thaw cycles, are reducing syrup harvests in New England. Maple syrup harvesters, or tappers, are starting their harvests nearly a month ahead of the previous generation. In contrast, Canada has seen a tripling of its maple syrup production. Though some of Canada's production increase may be due to economic incentives and changing markets, climate change is certainly favoring syrup production in Canada while hindering it in the United States. As with many other situations, other factors exacerbate the problem. The syrup harvest is being further reduced by the pear thrip, an insect pest introduced from Europe that causes widespread damage to maple trees. This insect is thriving in the increasingly moderate climate.

Climate impacts on biodiversity will affect people across a wide variety of ecosystems. For many coastal people in the tropics, coral reefs provide food and building material and support ecotourism, all of which can fail if coral bleaching takes its toll. In grazing lands, climate change will favor invasive species and increase fire frequency, which threatens the livelihoods of herders. Climate change has been

Rustan Swenson and Tim Hescock tapping maple trees at the Vermont Trade Winds Farm in Shoreham. Maple syrup can only be collected when temperatures cycle between temperatures that are above freezing during the day and below freezing at night. As spring temperatures continue to warm, the tapping season is shifting and becoming shorter in some areas.
© JOSHUA WOLFE

OPPOSITE: Coral bleaching off the Mariana Islands in the western Pacific Ocean.
DAVID BURDICK

associated with diseases that affect oysters and the toxic algal blooms that have adverse effects on coastal fisheries and the people and fishermen who depend on these sources of seafood.

Finally, climate change favors several disease organisms. It exacerbates health problems due to malnutrition and the impact of heat stress on cardiovascular disease and respiratory illnesses. Estimates of climate-related impacts on human health compiled by the World Health Organization stand at 150,000 deaths annually, mostly because of the increases in diarrhea, malaria, and dengue fever attributable to increased temperatures and rainfall in some areas. However, climate effects on health are likely to be complex interactions of the physical, ecological, and social environment. As climate change increases, those impacts will as well, particularly for societies that already have scarce resources and are least able to adapt.

BIOSPHERIC MELTDOWN?

No single aspect of global change is more prevalent or difficult to address than biodiversity loss. Climate change is one the leading drivers of this loss in the rich variety of life. By 2050, nearly half of Earth's biodiversity is predicted to be lost, due primarily to habitat conversion as landscapes are changed from natural systems to managed systems.

If we think of nature as a place free from most human influences, nature may no longer exist. Even if one were to find a patch of nature that is remote and isolated, it is still experiencing elevated carbon dioxide and human-induced climate change. Forests become plantations, productive grasslands become croplands, arid grasslands become grazing lands; even coastal systems can be converted to aquaculture. The surface area of the globe that is urban and suburban or paved with roads is continually increasing. The redirection of water from natural systems to managed systems by dams, dikes, canals, and dredging is also increasing.

Many other factors increase the stress on biodiversity: habitat conversion, the spread of exotic and harmful invasive species, unsustainable extraction of species such as fishes, unregulated clear-cutting of tropical forests, and air and water pollution. Elevated carbon dioxide and the large amounts of nitrogen being deposited in ecosystems from industrial and agricultural activities additionally put pressure on different parts of the ecosystem. Climate change is only one of several drivers, but it makes matters much worse and ranks among the top agents of change in the biosphere.

A good example is the introduction of kudzu in the southern United States to prevent soil erosion and for ornamental purposes. Unfortunately, it has spread out

Kudzu growth in Melville, Long Island. An invasive species, kudzu was initially imported from Japan to the southern United States for ornamental gardening and agricultural purposes, but it rapidly became a pest due to a lack of grazers and its fast growth. In the last ten years, it has moved inexorably farther north as temperatures have warmed, allowing it to reproduce in previously inhospitable areas.

© JOSHUA WOLFE

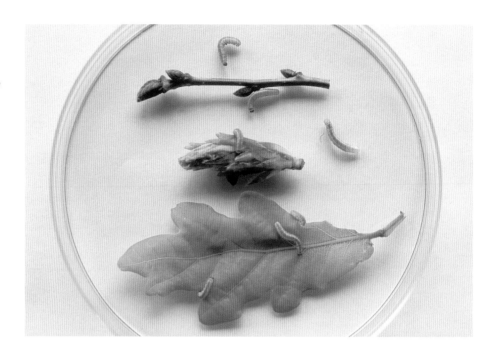

Winter moths (*Operophtera brumata*) at the Netherlands Institute of Ecology in Warnsborn. The caterpillars do best when they hatch at the same time as the bud burst of oak leaves in the spring. As spring has been coming earlier, both bud bursts and egg hatching have advanced as well. However, the egg hatching has advanced more than the bud bursts, leaving the caterpillars without sufficient food when they hatch. © PETER ESSICK

of control (over 28,000 square kilometers [7 million acres] are now infested), and its range is expanding northward, possibly in response to warmer temperatures.

Large numbers of species are shifting where they live because of climate change. Invasive species are becoming more problematic as climate change favors their success over native species, many of which are endangered. Climate change is either leaving them no place to move to or, especially if they are sedentary, changing too fast for them to adapt and survive. An estimate of potential rates of extinction based on the shrinkage of climate niches implies that 15 to 37 percent of the 1,103 species examined will be on the road to extinction by 2050. Based on this study, some have speculated that perhaps a million or more of Earth's species could face similar fates if the study's models are correct.

Climate change impacts are often portrayed as the tragedy of innocent species whose homes and interactions are being disrupted by the thoughtless actions of humans. In addition to the tragedy of extinction, we must look to the degradation or loss of the ecosystem services species provide. Recall that Earth's climate is fully capable of assuming alternative configurations, ranging from being ice-free to being covered from pole to pole with ice, and that biodiversity plays a major role in controlling which of these states Earth occupies. We should therefore have an enhanced appreciation of polar bears, penguins, corals, trees, migratory birds, and the remaining species for their role in stabilizing the biosphere.

Scientists have not had the time or resources to work out all the details or construct a massive blueprint of planet Earth's biosphere. We know crudely how it works, but not the role every species plays in its function. We do know enough to understand that species play important roles in climate processes. For instance, we know how much light is reflected by tundra vegetation and how much carbon dioxide is sequestered and stored by forests. Conversely, fire, a natural part of the nutrient cycle in ecosystems, can release carbon dioxide back into the atmosphere. Similarly, key chemicals and aerosols, which play important, though complex and still poorly understood roles in climate, are produced by plants, animals, and micro-organisms, particularly marine plankton (see Chapter 6). Finally, approximately half of the rain in the Amazon basin comes from the recycling of water by trees as they draw up groundwater through their roots and put it back in the atmosphere by the evaporation of water from their leaves. This one cycle alone is a clear example of how climate affects biodiversity and biodiversity affects climate. If significant tracts of rainforest are replaced with a monoculture of soybean, eucalyptus, or coffee, the amount of rain in Amazonia will likely drop dramatically.

Unlike biodiversity loss due to agriculture, fishing, or forestry, the effects of climate change on biodiversity are much more haphazard. When we clear-cut or restore a forest, or when we overfish or protect a fishery, these are deliberate acts almost always made in the interest of improving human well-being by securing the resources needed or desired by a growing population. As with most environmental problems, undesirable consequences may not be intentional or something we could have envisioned.

The evidence is overwhelming: the biosphere is changing. The order and structure of its biosphere's components, namely its species, are being reconfigured. We face a potential biospheric meltdown.

PART II

DIAGNOSIS

CHAPTER 6

CLIMATE DRIVERS

Tim Hall

Human beings are now carrying out a large scale geophysical experiment of a kind that could not have happened in the past nor be reproduced in the future.

—Roger Revelle

Climate is not static. It varies regionally and globally on time spans of decades to millions of years. Millions of years ago, dinosaurs lived on an Earth in which present-day Arctic regions were replete with subtropical trees and swamps. In the depths of the last ice age 20,000 years ago, much of now-temperate North America was covered with a mile-thick layer of ice, and the global surface air temperature was about 5°C (9°F) lower than today. As the Earth came out of the last ice age 11,000 years ago, it briefly—but suddenly—dipped back into a cold period that lasted a thousand years. From the tenth to the thirteenth century, coastal regions around the North Atlantic were warm enough to allow the Vikings to establish farming settlements on Greenland that were later abandoned when colder climate arrived in the fifteenth century. In the 1930s, the American Great Plains suffered a severe drought, colloquially known as the Dust Bowl, that devastated local agriculture.

In Earth's deep past millions of years ago, many factors that shape climate were different. The Earth's atmosphere contained different ratios of gases than it does today. The shape and location of the continents were changed. Variations in the Earth's rotation and orbit that occur in cycles over tens of thousands of years changed the seasonal amount of solar energy reaching the surface.

The Earth is currently warming. Since 1900 the globally averaged surface air temperature, determined from measurements at many meteorological stations, has increased about 0.8°C (1.4°F), with more than half of this increase occurring since

The number of climate drivers is large and sometimes difficult to characterize. This image, taken by a satellite in July 2002, shows a whole range of effects: ship tracks causing enhanced cloudiness in their wake (upper left), smoke from fires in Oregon increasing reflectivity over land and changing the thickness and color of clouds off the coast, and air pollution haze in the San Francisco Bay area. NASA

1970. Some regions, such as the Arctic, have warmed by greater amounts, while other regions, such as southern South America, have warmed little. Climatologists, who study the causes of climate change, have looked into many natural causes, some more plausible than others, put forth to explain the current warming. These include solar variability (the variation in the amount of sunlight that reaches Earth), cosmic rays (high-energy particles from outside the solar system hypothesized to affect clouds), and cycles in atmospheric composition or ocean circulation. All of these potential causes cannot explain the amount of Earth's warming since 1970. The one possible cause that has stood up to all tests based on direct observations is the increase of certain gases in the atmosphere in response to human industrial emissions.

SOLAR RADIATION AND GREENHOUSE GASES

Since John Tyndall first demonstrated it in the nineteenth century, we have known that some gases in the atmosphere act metaphorically like blankets, making it harder for the heat at Earth's surface to escape directly to space. These greenhouse gases include carbon dioxide (CO_2) as well as many others both natural and human-made, such as nitrous oxide (N_2O), methane (CH_4), ozone (O_3), and the ozone-destroying chlorofluorocarbons (CFCs). By far the most abundant greenhouse gas is water vapor (H_2O).

These and other atmospheric gases interact with light radiation in complex ways that depend on the energy of the light itself. Since light can be thought of as a wave, the energy of the light is inversely related to the wavelength. The Sun's light has a range of energy levels, from low-energy, longwave infrared to intermediate-energy visible light (that our eyes have evolved to detect) to high-energy, shortwave ultraviolet. Of these, visible light is the most abundant arriving at Earth. The ground and ocean warm up by absorbing sunlight, and some of this energy emanates back into the atmosphere as longwave infrared radiation.

What is the impact of light at these different energies striking the Earth's atmosphere? Some simple gas molecules (such as ozone) have internal components that vibrate easily in response to ultraviolet radiation. This means that the ultraviolet light that comes in is mostly absorbed by the ozone layer. Other, more complex molecules vibrate and rotate as a whole unit in response to longer wave infrared radiation. It turns out that gas composed of molecules with three or more atoms vibrate particularly well when struck. Think of them as bigger targets. Meanwhile, simpler gases whose molecules have only two atoms—such as the most common gases, oxygen (O_2) and nitrogen (N_2)—don't vibrate at all at these energies. No atmospheric gases interact with visible light to any significant extent, which is why we can see the stars. By contrast, the greenhouse gases interact significantly with the infrared heat radiation emitted by the Earth's surface.

Greenhouse gases could equally well be called blanket gases. Imagine yourself wrapped in a sleeping bag under a star-filled sky on a cold night. Your body generates heat, and the sleeping bag helps to keep that heat near your body. Similarly, the atmosphere helps to keep the heat radiating from the Earth near the surface. The analogy isn't perfect, however, because Earth doesn't generate a significant amount of heat on its own. The heat comes from the Sun. The magic of these particular gases is that they allow the Sun's energy, in the form of visible light, to pass through to the surface, but they inhibit the escape of the heat to space.

Infrared from the surface is absorbed by the greenhouse gases, heating up the atmosphere. The gases themselves radiate in all directions, including back to

Metaphors don't usually photograph well, but the blanketing effect of greenhouse gases is real enough. © JOSHUA WOLFE

the ground and off into space, and this bouncing around of the heat energy keeps the surface warmer than it would be otherwise. Paradoxically, high up in the stratosphere, where the ultraviolet absorbed by the ozone layer keeps things warm, the energy emitted by CO_2 actually serves to cool that region.

Not all objects in the solar system have significant atmospheres. For example, Earth's Moon and Mercury have no detectable atmosphere. Determining the temperature at the surface of such a body involves the relatively simple task of balancing the incoming solar radiation with the outgoing heat radiation at the surface. In the absence of an atmosphere, Earth would be on average a chilly –18°C (0°F). Greenhouse gases absorbing and emitting radiation keep heat energy in the lower atmosphere for a longer period before it ultimately escapes to space, warming the surface to a far more comfortable 15°C (59°F) on average. This is the greenhouse effect.

GREENHOUSE GAS CHANGES

The industrial era, which began in the eighteenth century (say around 1750), is marked by clear signs of human activity changing the composition of the atmosphere. Over a century ago, the Swedish chemist Svante Arrhenius had theorized that industrial emissions of CO_2 would lead to increasing levels in the atmosphere and might

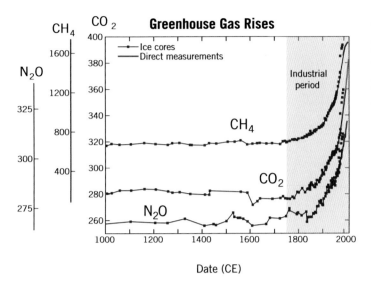

Greenhouse Gas Rises

CH$_4$ / CO$_2$ / N$_2$O

- Ice cores
- Direct measurements

Industrial period

CH$_4$

CO$_2$

N$_2$O

Date (CE)

For thousands of years, greenhouse gas concentrations were relatively stable, varying by a few percent from their mean levels. However, since the dawn of the industrial age around the year 1750, carbon dioxide (CO$_2$), methane (CH$_4$), and nitrous oxide (N$_2$O) concentrations have all increased dramatically. Note the different scales for each gas: CO$_2$ in parts per million, CH$_4$ and N$_2$O in parts per billion. DATA FROM THE NATIONAL CLIMATIC DATA CENTER

increase global temperatures. However, for many years it was widely, and mistakenly, thought that the ocean would absorb nearly all industrial CO$_2$. Not until the work of Roger Revelle and Hans Suess in 1957 was it appreciated that the ocean, although still taking up a significant fraction of CO$_2$, had a limited capacity due to peculiarities of carbon chemistry in seawater. Only a few years later, the careful, frequent measurements made by Charles Keeling on top of Hawaii's Mauna Loa 3,000 meters (2 miles) above sea level showed that levels of CO$_2$ in the atmosphere were, in fact, rising significantly.

As of 2008, CO$_2$ is present in the atmosphere at a level of about 384 parts per million (ppm) of air. This may not sound like much, but it is more than 36 percent greater than the 280 ppm concentration at the dawn of the industrial era, and more than the atmosphere has experienced for at least 800,000 years. Remember, human civilization began only about 10,000 years ago, and *Homo sapiens* have only been around for 150,000 years.

Other greenhouse gases also have risen substantially in the industrial era. Some are new gases that were previously absent from the atmosphere, while for others, human activity has boosted concentrations above their natural level. Methane is a naturally occurring atmospheric gas produced mainly by bacteria that decompose plant and animal matter without the presence of oxygen, usually in very wet environments such as swamps or wetlands. Methane's human-related sources come from agriculture (particularly rice cultivation), livestock (particularly cows and sheep), landfills, mining, and leakage from pipelines. Total methane emissions have more than doubled. Current concentrations of 1.8 ppm are 150 percent higher than natural background conditions (around 0.7 ppm). This is much less than CO$_2$ in total amount, but molecule for molecule, methane is about twenty times more effective as a greenhouse gas, although it only stays in the atmosphere for around a decade. In contrast to the currently increasing amount of CO$_2$, methane concentration in the atmosphere essentially stopped growing from 2000 onward. The detailed reasons remain obscure, but it is likely due to a reduction in sources from Eastern Europe. Enhanced emissions of methane into the atmosphere still take place, but apparently they have come into rough balance with the chemical breakdown in the atmosphere.

Nitrous oxide, or laughing gas, has increased 18 percent, from 0.27 to 0.32 ppm, over the same period. This increase stems chiefly from the rise of industrialized farming, which uses heavy doses of nitrogen-rich fertilizer.

Chlorofluorocarbons have no natural sources and were absent from the atmosphere before their invention in the 1930s. They were used as refrigerants and fire suppressors because of their remarkable stability. Concentrations increased exponentially until the 1980s when the effect they were having on stratospheric ozone was discovered. As a result of international treaties combating ozone depletion, their concentration has stabilized and will soon start to decline. As with methane, although the absolute concentrations of CFCs are much smaller than CO_2, they are much more powerful greenhouse gases.

FORCING CLIMATE

We can view increases in greenhouse gases as a thickening of the blanket that keeps the surface warm. This "thickening" can be quantified using the radiative forcing concept. Anything that changes the energy balance in the atmosphere can be characterized by how much it changes the radiation flux at the top of the atmosphere. Greenhouse gases are a positive forcing, as they prevent longwave energy from escaping into space, implying that more energy is coming in from the Sun than is going out. We can estimate the amount of radiative forcing from all the different factors to get a total net forcing. And that net amount is what controls the response of the climate to the forcings.

Greenhouse gases are by no means the only radiative forcing on climate. Changes in the Sun's intensity are another obvious example. Most well known is the eleven-year sunspot cycle, but there may be longer-term changes as well. Curiously, the Sun is slightly brighter when there are lots of sunspots because compensating bright areas surround them. Other radiative forcings change the amount of sunlight that is reflected from the Earth's surface. For instance, paved, roofed, or farmed surfaces absorb less energy and reflect more light than natural vegetation.

Radiative forcings are usually expressed as an energy flow—the amount of energy passing through an area over a particular length of time. Typically, the units used for energy flow are watts per square meter (W/m^2). A typical newspaper fully opened is about a quarter of a square meter, and a watt is simply a measure of the amount of energy use. For instance, a Christmas tree light is a 1-watt bulb. So 1 W/m^2 is equivalent to the energy flow associated with a 1-watt bulb on every square meter of the Earth's surface. The radiative forcing for a doubling of CO_2 is often used as a reference forcing and is roughly 4 W/m^2. The value for the 36-percent growth of atmospheric CO_2 so far over the industrial era is about 1.5 W/m^2. All other recent greenhouse gas changes combined (from CH_4, N_2O, and CFCs) amount to about another 1.2 W/m^2. For comparison, the net increase in the Sun's intensity

Air pollution levels in China have to be seen to be believed. The aerosol haze evident in this image is typical for Beijing, and may exceed the conditions seen even at the peak of the London smog in the 1950s. The sources of pollution are mainly the burning of sulfurous coal and the exhaust from cars and trucks in the surrounding region. GAVIN SCHMIDT

over the industrial era has a magnitude of between 0.1 and 0.3 W/m^2.

Negative radiative forcings, both natural and anthropogenic, cool the climate. They affect the radiative energy budget generally by scattering or reflecting incoming sunlight before it hits the surface. Volcanoes can emit large amounts of sulfate aerosols into the upper atmosphere, resulting in a negative forcing that may be larger than 4 W/m^2 in magnitude, though the effect is short-lived. For instance, the most recent climatically significant eruption was the June 1991 eruption of Mount Pinatubo, in the Philippines, which resulted in a temporary negative forcing of about 4 W/m^2. Industrial activity also has put aerosols into the atmosphere (discussed in further detail below), but with a likely net negative effect of about 1 W/m^2. Change in land use also has forced climate, by altering the albedo (reflectivity) of Earth's surface to incoming sunlight and through the altered cycling of water. For example, a field of wheat is more reflective than a forest, so a change from forest to field represents a negative forcing. Globally, the radiative forcing of land-use change is estimated to be a decrease of 0.1 to 0.2 W/m^2, though locally its magnitude can be much larger, and either negative or positive due to the complexity of the interactions.

Ozone changes have complex radiative forcings. Near the surface, "bad" ozone has increased because of urban smog and air pollution, giving roughly a positive 0.4 W/m^2 forcing, although the effect varies regionally. On the other hand, depletion of "good" ozone in the upper atmosphere has resulted in a small negative forcing, of about 0.1 W/m^2.

To summarize, the total net positive forcing from 1750 to today is about 1.6 W/m^2, coincidentally about the same value as for CO_2 on its own. Of the main greenhouse gases, CO_2 has provided the major greenhouse gas forcing (56 percent), with the remainder derived from methane (16 percent), nitrous oxide (5 percent), CFCs (11 percent), and low-level ozone (12 percent).

CLIMATE RESPONSE

How does climate respond to the radiative forcings? Perform a thought experiment: Start with a quiet Earth in energy balance (the amount of energy coming in is the same as the amount leaving), then instantaneously increase the radiative forcing, for example by doubling the amount of CO_2 in the atmosphere, and keep it at that level indefinitely. Now more energy is coming in than is leaving, and the planet will begin to warm. Eventually the atmosphere and ocean will come into a new balance at a higher temperature. The change in global mean surface air temperature is called the climate sensitivity, and it is a convenient tool to summarize the impact of a forcing.

Global climate models warm about 3°C (5°F) with doubling of CO_2, with a range of about plus or minus 1°C. These modeled climate sensitivities are corroborated by the sensitivity inferred from observations and the analysis of Earth's past. For example, in the depths of the last ice age roughly 20,000 years ago, the surface was more reflective due to the much larger ice sheet coverage and reduced vegetation, and the atmosphere contained less CO_2 and other greenhouse gases and more dust—all of which are cooling effects. Consequently, radiative forcing was negative compared to the recent preindustrial past. We can estimate the values of that forcing by combining our knowledge of greenhouse gases and albedo to calculate an approximate 7 W/m² decrease. The global air temperature was consequently lower, by about 5°C (9°F). The resulting estimate of the climate sensitivity, 5°C for 7 W/m² (equivalent to 2.9°C [5°F] per doubling of CO_2), closely agrees with climate sensitivities seen in models.

Earth's climate system is not a laboratory in which controlled experiments can be performed. Only in a computer model can one, for example, increase the concentration of CO_2 by a fixed amount, holding all other climate variables constant, and observe the isolated effect on air temperature. In reality, many radiative forcings are changing simultaneously. In addition, their effects on climate interact. The change in air temperature in response to one radiative forcing induces another radiative forcing, the impact of which may be to reinforce or counter the original temperature change. (Reinforcement is synonymous with positive feedback, and countering is synonymous with negative feedback. Remember, in a climate context, positive feedback does not imply a runaway effect whose magnitude continues to grow without bound, but merely an amplification of the initial change.) The climate sensitivity summarizes the total temperature change due to a myriad of feedbacks in physical, chemical, and biological processes in response to a doubled-CO_2 radiative forcing.

The single most important climate feedback involves water vapor, which is a gas with an enormous natural source: the ocean. We know from everyday experience that the amount of water vapor that can remain in the air without condensing is a strong function of temperature. A hot, clear day can be humid, but a cold, clear day can only be dry. If on a hot, humid day the temperature abruptly drops, the water vapor condenses and it rains. The leftover puddles evaporate quickly if the air warms up again.

As Earth warms, the air has the capacity to hold more water as vapor. Water vapor, in its role as a greenhouse gas, absorbs additional heat radiating from the surface, further warming the atmosphere and surface. This mechanism works for all causes of warming, whether from CO_2 or the Sun, and is always a significant effect. The water-vapor feedback roughly doubles the effect seen in experiments with climate models and analysis of the climate response to volcanic eruptions.

Why should we be concerned about the greenhouse warming of CO_2 when water vapor is a greenhouse gas with much more prevalent natural sources? Do climate scientists ignore the dominant "natural" warming of water vapor when discussing anthropogenic climate change? The answer is, of course, no. Climate scientists do not ignore water vapor, and they recognize its dominant greenhouse effect. However, water vapor is not a radiative forcing external to the atmosphere-ocean system. The industrial sources pump out CO_2 gas regardless of the surface air temperature. The atmospheric abundance of CO_2 depends only weakly on temperature, at least at present, via small interactions of the carbon cycle with climate (see the carbon cycle discussion below).

Water vapor, by contrast, is highly dependent on air temperature. Globally, the relative humidity of the atmosphere—the water-vapor concentration *relative* to the water-vapor concentration at which it would condense—stays relatively constant over wide temperature variations. Add water vapor to the air at constant temperature, and it will condense and rain, returning approximately to the same humidity. Remove it, and within days new water will evaporate from the ocean to fill its place. Similarly, increase the temperature, and soon enough water will enter the atmosphere to reach the new concentration required to keep relative humidity roughly constant. Reduce the temperature, and almost immediately the water will precipitate until the new level is reached. Water vapor is, therefore, a slave to temperature and is better thought of as a feedback, not a forcing, whose amplifying effects are crucial. These effects are represented in all climate models.

Another important climate feedback works through the high reflectivity of ice. This positive feedback has played a key role in the warming of the Arctic, which has outpaced the Earth as a whole (as discussed in Chapter 2). More generally, the ice-reflectivity and other feedbacks cause polar regions to have a substantially higher climate sensitivity than nonpolar regions, a feature known as polar amplification.

A third set of feedbacks involves clouds. Clouds absorb outgoing heat radiation and thus add to the greenhouse effect as well, but they are also highly reflective, cooling the planet by reflecting solar radiation back to space. The net effect of

clouds depends on their altitude. Low, thick clouds reflect more sunlight back to space than they absorb, cooling the surface. For high, thin clouds the balance is the opposite—they also reflect sunlight, but due to the cold temperatures aloft, they are particularly effective at absorbing infrared. Changes in cloud amount, type, and distribution can, therefore, either amplify or counter greenhouse-induced warming. Clouds might change because temperature is changing, water vapor content is changing, winds are changing, or the abundance of aerosols that can nucleate cloud droplets is changing. These effects and interactions happen at very small scales and with very complex physics. Thus, simulating cloud feedbacks is a great challenge. In most models, the high-cloud positive feedback dominates, so that, globally, clouds amplify greenhouse-induced warming. Regionally, models and observational analysis show that both positive and negative cloud feedbacks occur.

Cloud tops as viewed from the space shuttle. The cumulus clouds are the tops of large convective systems, and the smoother stratus clouds are the thinner offshoots. Each type of cloud has different optical properties, and changes in their distribution can profoundly affect the atmosphere.
NASA

THE CARBON CYCLE

Carbon is so important to global climate that its processing by the climate system deserves a detailed discussion. Consider a molecule of CO_2 or methane released

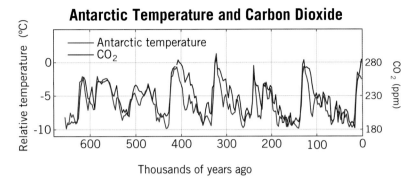

Antarctic Temperature and Carbon Dioxide

Over eight glacial cycles in 650,000 years, global temperature and the amount of CO_2 in the atmosphere have gone hand in hand. When temperatures are high, so are CO_2 amounts, and vice versa. This obvious connection is part of a coupled system in which changes in climate affect CO_2 levels, and CO_2 levels also change climate. The pacing of these cycles is set by variations in the Earth's orbit, but their magnitude is strongly affected by greenhouse gas changes and the waxing and waning of the ice sheets. DATA FROM THE NATIONAL CLIMATIC DATA CENTER

into the atmosphere. How long does it stay there? Where does the molecule go after it leaves, and does it come back? How does chemistry affect the carbon? These questions fall in the domain of carbon cycle science, which seeks to understand the cycling of carbon, both natural and human-produced, through its many reservoirs (the components of the climate system where we find carbon).

The natural carbon cycle is complex. Different reservoirs contain vastly different amounts of carbon, evolving on vastly different timescales. Globally, almost all carbon is contained in the solid earth. The exchange of this reservoir with the atmosphere and ocean by weathering, erosion, and sedimentation occurs over millions of years. Over centuries, however, the solid earth is stable, and the dominant exchanges of atmospheric carbon occur between the land and marine vegetation and the ocean. Carbon dioxide is highly soluble in seawater, and the ocean holds fifty times more carbon than the atmosphere. Plants and algae consume CO_2 during photosynthesis, converting the carbon to organic material. Animals consume plants. Plants and animals respire, returning CO_2 to the atmosphere. Plants and animals die and decay, returning some of the carbon to the atmosphere and the sea floor.

The components of the carbon cycle are never truly in balance. Their exchanges both evolve in response to climate and cause climate to evolve in complex, subtle, ways. For example, roughly 40 percent less CO_2 is found in the atmosphere at the depth of an ice age than during a warm period. Many processes that could drive this variation have been proposed. Although their contributions are uncertain, the increased CO_2 solubility in cooler seawater is likely key, along with changes in ocean productivity. Ice age CO_2 variations, in turn, play a role in forcing climate, due to their greenhouse effect discussed above. That is, the climate impacted CO_2 levels at the same time that CO_2 was altering climate. Analyses of ice age climate indicate that just under half the cooling was due to reduced greenhouse forcing.

Despite these large natural CO_2 variations, atmospheric CO_2 remained relatively stable over the 12,000 years from the end of the last ice age to the dawn of the industrial era, varying between 260 and 280 ppm. Methane, too, was stable during this period, varying from 0.6 to 0.7 ppm. These trace-gas concentrations are well known from analyzing air bubbles trapped in ancient snowfall.

This relative stability came to an abrupt end with the onset of the industrial era. At that point, we started transferring to the atmosphere carbon that had been stored in underground reservoirs for millions of years. These modern increases have occurred in a geologic blink of the eye, dwarfing the rate of increase coming out of the last ice age. Plotted on the same graph as the ice age change, the industrial era increases look like vertical lines.

Anthropogenic carbon is such an abrupt perturbation to the natural carbon cycle that many components of climate have not yet had time to respond and interact with the new carbon. It is a great challenge in climate change research to quantify the terms of the "budget" of anthropogenic carbon—that is, where the carbon emitted by human activities is ending up.

The amount of anthropogenic carbon in the atmosphere is the best-known component of the perturbed carbon cycle. Carbon dioxide is chemically inert in most of the atmosphere, so no chemical changes need to be considered. Winds and turbulence mix the atmosphere quickly compared to the rate of increase of CO_2, so CO_2 concentrations are relatively uniform, varying by only a few parts per million between the Southern and the Northern hemispheres. Consequently,

The industrial era really got started in the eighteenth century with the large-scale use of coal in steam engines and for power generation. Even today, coal is still the largest energy supplier for human activities, especially in Inner Mongolia, China, where this truck is being loaded up. © GARY BRAASCH

A coal-burning plant in Conesville, Ohio. The stack with the white smoke (second from left) is using lime to reduce sulfur emissions and thereby reduce downwind air pollution and acid rain.
© PETER ESSICK

a sparse network of surface monitoring stations is sufficient to estimate the atmospheric abundance. For CO_2 concentrations prior to 1950, we rely on analysis of air bubbles trapped in ice core samples. Taken together, these data show that CO_2 has increased steadily over the industrial era, and that the rate of increase has itself increased. Roughly 50 percent of the 100 ppm increase since 1850 has occurred in the past thirty years, and the growth rate of 2.0 ppm per year (about 0.5 percent per year) in the period from 1998 to 2007 is the fastest on record. The accumulation in the atmosphere in this period is about 4 gigatons of carbon (GtC; 10^9 metric tons) per year.

The next best-known term in the anthropogenic carbon budget is the industrial injection, due primarily to fossil fuel burning. Based on records of activity in the energy and transportation sectors of the economy and on correlations with economic growth, about 7 GtC per year were injected into the atmosphere from 1995 to 2005. Additionally, roughly 1 GtC per year was injected from intentional burning of (mostly tropical) forests to expand agricultural lands, although this figure is less certain.

Notice that the rate of accumulation in the atmosphere is only half the rate of injection. This airborne fraction has remained surprisingly steady over the industrial period, staying near 50 percent. Quantifying in detail where the rest of the carbon has gone (the "sink") is one of the aims of carbon cycle science.

Typically, the sink is partitioned into land and ocean. Of the two, ocean carbon uptake is best, though not perfectly, constrained. Carbon dioxide dissolves in seawater, and the oceans have great capacity to absorb anthropogenic carbon (though it is limited by the buffering effect discussed in Chapter 3). There are three major reasons why estimating anthropogenic carbon in the ocean is more difficult than in the atmosphere: (1) the magnitude of the oceanic carbon perturbation is much smaller (about 1 percent), due to the larger natural amount of carbon in the ocean; (2) carbon in seawater is not inert, but rather undergoes complex biochemical transformations; and (3) due to the ocean's slow circulation and mixing, much of the deep ocean has not yet been contaminated by anthropogenic carbon, so that the anthropogenic carbon distribution is far from uniform.

Despite these difficulties, many estimates of ocean carbon uptake have been made, using a variety of independent techniques. They can be based on numerical models of ocean circulation, along with representations of carbon chemistry; or on the diagnosis of ocean transport using other, simpler, trace constituents; or on the direct measurements of dissolved carbon over time. All of these techniques conclude that an uptake of 1.5 to 2.5 GtC per year has taken place over the past decade.

Going into the atmosphere there is 7 GtC per year from industrial activity plus 1 GtC per year from forest burning, of which 4 GtC per year accumulates in the atmosphere and 2 GtC per year enters the ocean. This leaves 2 GtC per year unaccounted for (give or take at least 1 GtC per year, given the uncertainty in forest burning and ocean uptake). By process of elimination, this carbon must be taken up on land, likely by additional vegetation mass. Increased CO_2 in the atmosphere may accelerate photosynthesis, leading to new plant growth and thereby increasing carbon uptake. Many other factors could hamper this effect, however, such as limitations on nutrient and moisture availability. The uptake increase could also be compensated by an acceleration of respiration and release of carbon by microbes in soils as temperatures increase. It would be reassuring if estimates of changes in terrestrial biomass corroborated the 2 GtC per year uptake implied by the rest of the carbon budget, but precise global estimates are not yet possible due to limited measurements and the extreme "patchiness" of ecosystems.

In a warmer climate with different rainfall patterns, plant types will wax and wane in number and move into new terrain, changing patterns and possibly the magnitude of land-atmosphere CO_2 exchange. Melting permafrost in high latitudes may release CO_2 and methane currently stored in the soil. Possible feedbacks abound in the ocean as well. Carbon dioxide is less soluble in warmer water. In addition, the peculiarities of inorganic carbon chemistry in seawater are such that higher amounts of dissolved CO_2 impede further uptake. Other things being equal, the ocean's ability to absorb anthropogenic carbon will decline as temperatures and atmospheric CO_2 rise. Additionally, warmer upper waters in the ocean increase vertical stability (warm over cold is stable), impeding the upwelling from depth of nutrients required for surface-water biological activity. This could further reduce photosynthetic uptake of CO_2 in the oceans.

Each of these effects will lead to changes in the airborne fraction and greenhouse gas concentrations, potentially leading to enhanced future warming if the net effect is positive (as is inferred from the ice core records and as is generally expected).

There are two important points to add. First, because of the slowness of ocean uptake, CO_2 concentrations will take decades and centuries to come back down, even if we stop emitting CO_2 immediately. Roughly 15 percent of the carbon we emit will still be in the atmosphere five hundred years from now. Second, simply stabilizing CO_2 at the concentration it is now requires a reduction in emissions of around 60 percent, and because of climate feedbacks, the reduction may need to be closer to 80 percent in the long term.

AEROSOLS

Aerosols constitute another radiative forcing agent deserving of detailed discussion. The haze that often limits visibility in urban areas is due to aerosols that are either directly emitted as pollutants or produced from other pollutants. Other aerosols have natural sources, such as wind-blown soil and dust, organic aerosols from vegetation, volcanic sulfur, and sea salt evaporated from the ocean. Anthropogenic aerosols include sulfate (derived from the sulfur dioxide produced when burning fossil fuels), nitrates, and black carbon (soot) from incomplete combustion and deliberate forest burning for land clearance. Aerosol pollution in industrial regions often causes local human health problems such as bronchitis and asthma. Regionally, aerosols also are implicated in other environmental problems, such as acid rain. Many areas in the Northern Hemisphere have had significant decreases in sunlight at the ground since the 1950s. This "global dimming" is mostly associated with increasing aerosols combined with their affect on clouds.

Aerosols play a crucial role in perturbing the radiative energy budget, both directly, by scattering and absorbing solar radiation, and indirectly, by altering the properties of clouds. The radiative properties of aerosols are complex and depend on size, composition, and height in the atmosphere. Large volcanoes in tropical latitudes can inject huge amounts of sulfate aerosols into the stratosphere, where they slowly drift down over a couple of years. The sulfate aerosols are white and so reflect incoming solar radiation, cooling the Earth's surface. For instance, in the years following the Mount Pinatubo eruption, the global mean surface air temperature cooled by a maximum of about 0.5°C (1°F). Anthropogenic sulfate aerosols also cool the surface, though because they are easily washed out of the lower atmosphere by rain, they have a much shorter residence time of only a few weeks.

Other aerosols have very different effects. In particular, black carbon is an umbrella term that refers to the products of incomplete combustion and includes charcoal, chars, and soot. The small particles are carried into the sky by the hot air from combustion and eventually return to earth, washed out by rainfall or simply settling like dust. The climate change concerns are twofold: in the atmosphere they darken the sky, and when they fall on snow and ice they darken the surface. Both effects lead to decreasing reflectivity and hence a warmer planet. Scientists have measured the increasing black carbon content of snow in the Arctic and Greenland, but have a low level of understanding regarding the magnitude of the effect. Polluted air in China and India is so evident that people have named it the Asian Brown Cloud, much of which is made up of soot aerosols.

The aerosol indirect effect works by modifying many characteristics of clouds, including water-droplet size and number, and the lifetime and height of clouds. Clouds themselves are crucially important in blocking sunlight from reaching the ground and in blocking heat from escaping to space, depending on their type. Changes in cloud properties therefore can have a large impact on the radiative budget. Unfortunately, due to the complexity of clouds and aerosol interactions, there is a great deal of uncertainty about the size of the indirect cloud effect of aerosols. The overall effect of anthropogenic aerosols, however, is thought to be strongly cooling.

Ironically, although these pollutants cause other environmental and health problems, roughly twice as much warming would have taken place over the twen-

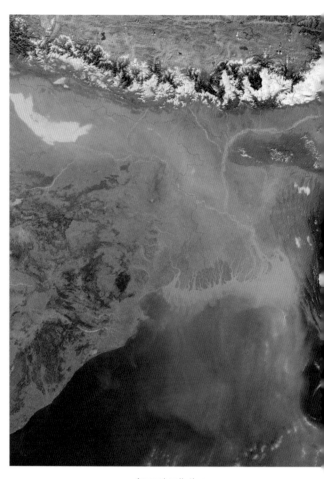

Aerosol pollution over northern India and Bangladesh as seen by the MODIS satellite instrument in 2001. The skies over the Tibetan plateau are very clear, indicating that the pollution is at a relatively low altitude and is being blocked by the Himalayas. The aerosol soup consists of sulfates, nitrates, organic and black carbon, and fly ash, mainly from industrial activity. In addition to the direct health impacts from breathing this air, there are likely impacts on rainfall and hydrology in the region. NASA

tieth century without them. The inclusion of anthropogenic aerosols as a radiative forcing is required for climate models to simulate realistically the observed surface air temperature changes. The mild reversal in mid-twentieth-century warming was due, in large part, to the growth in aerosols. As discussed in Chapter 4, the impacts of the aerosol changes may have been felt in the Sahel region and in Atlantic hurricane activity, enhancing the former and reducing the latter. Growth in aerosol emissions in the last few decades has slowed, particularly in the United States and Europe, where clean air legislation has led to significantly lower emissions from power plants. In Asia, however, aerosol emissions have grown rapidly, making their overall impact on climate in the last few years difficult to determine.

THERMAL INERTIA

We have discussed forcings and equilibrium response, but the climate is not presently in equilibrium; things are instead changing quickly, and this fact has important implications for the future. Consider that when you turn on your stove to boil a pot of water, the temperature in the water does not change instantaneously. As heat is added to the water from the burning gas, the water temperature gradually increases. Depending on the gas flow, the amount of water, and the composition of the pot, you might be able to drop in your pasta in ten minutes. This lag of the water-and-cooking-pot system is called thermal inertia: the larger the thermal inertia, the larger the lag.

Earth's climate system has considerable thermal inertia. As heat is added to the atmosphere, ocean, and land systems due to radiative forcing, the temperature gradually responds. Of the three components, the atmosphere has the lowest thermal inertia. If that were the only component of the climate system, its temperature would simply follow greenhouse gas concentrations with little or no delay. The other components, particularly the ocean, have much greater thermal inertia, due to the high heat capacity of water, the slow circulation of the ocean, and the inability of radiation at the sea surface to penetrate much more than a few tens of meters. Much of the water in the deep ocean has not been to the surface in centuries, and therefore has not had a chance to respond to anthropogenic forcing.

Experiments with ocean models indicate that, overall, the upper ocean takes several decades to equilibrate with an increase in heat input. The implications for climate change are important: if greenhouse gases in the atmosphere were stabilized now, the Earth's temperature would continue to increase, because the oceans are still responding to forcings of several decades ago. A certain amount of future warming is therefore already "in the pipeline." Estimates based on the observed rate of warming of oceans and confirmed by climate model simulations indicate that this guaranteed further warming will be about 0.6°C (1°F) over the next several decades.

ATTRIBUTION AND THE NATURE OF "PROOF"

Our climate is unequivocally warming, and it is very likely due to human activity—principally the emission of greenhouse gases. These two conclusions represent the synthesis of a vast amount of scientific work on climate and its recent evolution by thousands of researchers worldwide. That climate is warming is nearly irrefutable and is based on the many direct surface air measurements around the globe over the past century and collaborated by extensive Arctic ice melt, ocean warming, earlier spring onset, and longer growing seasons. That humans are likely responsible for the warming is a conclusion based on extensive testing of various possible explanations of global warming. The explanation for global warming in terms of industrial greenhouse gases is based on sound physical principles, all of which have stood up to numerous observational tests, unlike the many other proposed theories.

It is not possible in any field of natural science to absolutely prove that an explanation for a phenomenon is correct, only that one is incorrect or inadequate. Many explanations may be put forth to explain a series of observations or results from an experiment. The explanations, to be most useful, make distinct predictions about other observations. As additional observations become available, some explanations are shown to be incorrect, while others remain consistent with the data. Explanations may be modified in light of new data, in which case they must be tested against further data. The continual interplay between observational measurement and the development and modification of explanations comprises the scientific process.

No one, then, can prove conclusively that the observed global warming is due predominantly to anthropogenic greenhouse gases. However, given our present understanding of the climate system, the presence of increasing greenhouse gases constitutes a compelling explanation, or theory, of the observed warming. The theory of the anthropogenic role in global warming is an explanation in the same sense that the theory of gravity is an explanation for the mutual attraction of objects with mass, or that the theory of plate tectonics is an explanation for the distribution of the continents, or that the theory of evolution is an explanation for the development of biodiversity.

The theory of anthropogenic climate change has made predictions that have been borne out. Climate change theory has correctly predicted that sea levels should rise as ice melts and warmer seawater expands. Climate change theory has correctly predicted that the stratosphere should cool in response to increasing levels of greenhouse gases. Climate change theory has correctly predicted that warming in the Arctic would be enhanced due to ice melt and the subsequent energy absorption of exposed seawater. Climate change theory has correctly predicted that heat content would rise in the ocean, and that the land would warm more rapidly, all of which have been observed.

Where initial theories failed—for example, early climate model overestimates of twentieth-century warming—they have been modified to include other effects, such as the inclusion of aerosol changes. And the modified theory has subsequently made additional successful predictions—for example, the spatial distribution of warming, and the "global dimming" of sunlight near the ground in polluted regions.

The fundamental physics of the absorption and radiation of heat by greenhouse gases has been understood since the nineteenth century, and direct observations of the large increase in atmospheric greenhouse gas concentration have been made since the mid-twentieth century. These simple components of the theory, coupled with a few key feedbacks, are so compelling that, if a new theory of global warming were put forth, not only would it have to explain observed climate change over the industrial era, it would also have to explain why the climate was not responding to greenhouse gases in a way consistent with the known physics. No one has put forth any such theory to compete with anthropogenic influence.

Climate researchers have a simple reason to think that humans play a dominant role in warming. When all the known natural and anthropogenic forcings are used in an analysis of the industrial era, the observed temperature evolution, including its spatial variations and "fingerprints," can be matched reasonably well. When only the natural forcings are used, the temperature evolution cannot be reproduced, especially the rapid warming since 1970. Conclusions have not been reached simply because of a lack of imagination in thinking of other explanations. Indeed, climatologists have a long history of dreaming up theories to explain past changes in Earth's climate, with varying degrees of success. But when applied to current climate change, these theories are inadequate. For instance, explanations based on solar activity fail, because observed solar variations have shown no trend in recent decades, and the fingerprint of cooling in the stratosphere is inconsistent with known solar effects. In contrast, the explanation in terms of anthropogenic greenhouse gases has been shown to match a huge array of independently gathered data. Warming due to anthropogenic greenhouse gases is the best explanation for current trends; it has survived many tests, and it has no viable competitor.

Naomi Oreskes

THE SCIENTIFIC CONSENSUS ON CLIMATE CHANGE

Scientific experts have a consensus: human activities are changing Earth's climate. Many people think this is a recent accomplishment—that we finally know for sure after many years of uncertainty. In fact, the consensus has been around for a long time, and it came in two stages.

In the late 1970s, scientists realized that increased greenhouse gases would lead to global warming and the consequences could be grave. Early climate models suggested that if you doubled the amount of carbon dioxide in the atmosphere, average world temperatures would rise 2°C to 3°C (3°F to 5°F) and produce heat waves and droughts, rising sea level, and changes in the distribution of plants and animals. They also predicted that the effects of warming would be greatest at the poles. This was the first stage of consensus: a prediction that warming would occur.

When would this happen? Most experts thought twenty-five to fifty years. Others pointed out that "doubling" was no magic value, just a convenient level for discussion. If a 100 percent increase in carbon dioxide would lead to a 3°C rise in temperature, it stood to reason that a 30 percent increase might lead to a degree or so. In 1979, carbon dioxide had already risen about 20 percent, so some bold souls believed that changes were probably already occurring.

Still, most scientists thought that NASA scientist James Hansen was out on a limb when he declared in 1988 that global warming had arrived. He was, however, soon vindicated. In 1995, the Intergovernmental Panel on Climate Change concluded that the human effect on climate was "discernible." Temperatures were rising, spring was coming earlier, and the Arctic polar ice was melting, just as predicted. This was the second stage of the consensus: the predictions were coming true.

Yet, all along, a few people have challenged the science. They have claimed that there is no warming, or it is just natural variability, or it isn't necessarily bad and there's nothing we can do about it anyway. Lately, these challenges have taken a new form— arguing that "science is not about consensus."

The argument goes something like this: Science progresses through the bravery of lonely heroes (think Newton and Galileo) who challenge conventional wisdom. To emphasize consensus is to enshrine dogmatism over inquiry. Climate scientists who insist on the consensus view are suppressing alternative views for ideological reasons.

For these folks, climate contrarians are heroes, because they stand up for unpopular views in a sea of self-satisfied dogmatism. Some contrarians have compared themselves to Alfred Wegener, the German geophysicist who first developed the theory of continental drift only to have it rejected by the vast majority of earth scientists.

This challenge has the shoe on the wrong foot. It is true that scientific mavericks can trigger important changes in thinking, but until their ideas have been vetted they are just that—ideas. An idea becomes scientific only when it is supported by evidence, the evidence has been subjected to critical scrutiny, and a community of experts has concluded that the evidence supports the claim and there isn't any better explanation. When the arguments are done, consensus is what's left. Alfred Wegener is not a hero *because* his idea was rejected; he's a hero because in time his idea was seen to be *right*.

Robert Merton, a famous sociologist who specialized in the field of science, noted that a crucial characteristic of scientific inquiry is "organized skepticism," which means that all scientific claims must be deeply scrutinized. But his emphasis was as much on the organization as the skepticism: the scrutiny is done in a systematic way, through conferences and peer review. If a claim clears these bars, then scientists accept it and turn to other matters. The progress of science doesn't just require new ideas, it also requires the ability to settle a debate and move on.

There is a world of difference between being ahead of your time and being unwilling or unable to accept a view that has been vindicated by decades of scientific work. It is the difference between being innovative and being stubborn, between a maverick and a mule.

The correct analogy with climate contrarians is not Alfred Wegener, but Sir Harold Jeffreys, a brilliant geophysicist who rejected continental drift when it was first proposed in the 1920s and continued to do so long after the scientific community judged the evidence to be overwhelming. Over five decades, Sir Harold's objections scarcely changed, and he couldn't accept that—as smart as he was—on this issue he had backed the wrong horse.

The Alfred Wegener of climate science is Guy Callendar, who in the 1930s was the first to argue that temperatures were rising because of increased atmospheric carbon dioxide and more was in store. Callendar's claims were not immediately accepted; it took a lot more work for the evidence to be generally convincing and for important theoretical questions to be resolved. Similar stories could be told about relativity, quantum mechanics, and the origin of species by natural selection. None of these ideas were accepted overnight.

Science is intrinsically conservative, because the burden of proof is on the person who wants to dislodge accepted facts and theories, and it doesn't make sense to throw away hard-earned knowledge for any idea that just happens to come across the transom.

When agreement is forged on a new theory, individuals can dissent from the new consensus, *of course*, but unless they can come up with some new evidence or new arguments, their dissent becomes unproductive. There comes a point where disagreement degrades into denial. At that point, the scientific community stops paying attention, because further arguments without further evidence are a waste of time.

This is the case here. Back in the 1980s, dissenters raised legitimate questions about how climate models dealt with clouds, whether increased atmospheric carbon dioxide would produce more luxuriant plant growth that would slow atmospheric accumulation, and whether ocean absorption of carbon dioxide and heat might solve the problem. These questions—and many more—have been answered: neither clouds, nor plants, nor the oceans will save us from ourselves.

CHAPTER 7

STUDYING CLIMATE

Gavin Schmidt and Peter deMenocal

Write how the clouds are formed and how they dissolve and what it is that causes vapor to rise from the water of the Earth to the air . . .
—Leonardo da Vinci

The wind that blows is all that anybody knows.
—Henry David Thoreau

From October 1997 to September 1998, dozens of scientists and graduate students could be found drifting on the sea ice in the Beaufort Sea, north of the Canadian Arctic. As well as spending a cold and isolated winter thousands of miles from home, the scientists also had to be constantly ready with a rifle to scare off any hungry polar bears that happened to wander into view. Rather than simply being the unfortunate victims of a shipwreck, most had actually volunteered to be there. What motivated them? And why should we care?

The study of climate is one of the most complex and lively branches in all of Earth science. It is an amalgam of dozens of different fields: meteorology, oceanography, biology, chemistry, quantum physics, orbital mechanics, and ecology. As one might expect, the scientists studying climate are as diverse as climate itself.

Climate scientists can broadly be split into four overlapping groups according to discipline. Some study the physical processes in the current climate system, others look for indications of how and why climates were different in the past, others document the impacts of change today, and some bring all of these elements together so that they might be able to say something about tomorrow. Each branch contributes something unique to the mix, and our understanding of climate change and its impacts owes something to each of these subfields.

The launch of NASA's Terra satellite in December 1999. Terra is one in a constellation of five satellites in low-Earth orbit that follow each other in the same orbit, separated only by a few minutes. Measurements from Terra include the Earth's radiation and the aerosol amounts and concentrations of pollutants in the Earth's atmosphere. In particular, Terra hosts the MODIS instrument, which was used for a number of satellite images in this book. NASA

Two Canadian Coast Guard icebreakers unloading equipment to set up the SHEBA base camp on an ice floe in the Beaufort Sea in October 1997. The ice is 1.6 meters thick and floating on top of 2,000 meters of deep water. DON PEROVICH

OBSERVERS OF CLIMATE PROCESSES

The Earth's climate is a huge interconnected system that ranges from the ocean floor to 100 kilometers above the surface of the Earth; from the frigid wastes of Antarctica to the lush tropical rainforests of the Amazon; from the scale of an individual cloud droplet to ocean gyres thousands of kilometers across. Some of these places and processes are easily accessible. But most are not, so we need a large array of different techniques to probe them.

Climatology as a discipline exists because all of these scales are relevant. But most processes are still studied in a traditional reductionist way. Researchers have become experts in trying to slice and dice the climate into more manageable bite-size pieces for study. None of those pieces are complete, but each measurement gives a different perspective on the system as a whole, which can then be integrated into a coherent picture.

The information that is most accessible is from the surface, particularly on land, and is typified by the networks of data-gathering stations that have been developed to help in weather forecasting. These weather stations often consist of just a small screened box containing a thermometer and a barometer (for measuring pressure) with an anemometer (for measuring winds) on the side. These instruments feed their information directly to the national weather services in individual countries, which then forward this data to the central databases that are used by the weather-forecasting centers, such as the National Weather Service in the United States. Some of these stations have provided the longest records of instrumental data, which are some of the clearest measures of what has happened since about 1850—the instrumental period.

Long, coherent records are particularly prized; some individual records, such as at the Royal Observatory in Paris, go back four hundred years to the invention of the mercury thermometer. Since the records are often in Latin or in archaic scripts, graduate students in history often are needed to transcribe the old handwritten records, and this process is still ongoing.

An important point to note is that the needs of weather forecasters and climatologists are subtly different. Weather forecasters need the most accurate and up-to-date information at all times, whereas climatologists are interested in the statistics, such as the averages and variability of that data. Moving weather stations around, or updating their equipment, may help improve the weather forecast, but these changes can play havoc with analyses of the long-term climate trends by introducing subtle biases in the climate record. For instance, a weather station in the center of a city often records the effects of increasing urbanization over time, which is not

An original handwritten page by Giovanni Cassini from the Paris Observatory's weather logbook from the eighteenth century illustrates the difficulty of dealing with older sources of data. Both the handwriting and the meteorological conventions need deciphering. The second image is an example of the collated data showing the weather records from a crucial few days during the French Revolution. LEFT: PASCAL YIOU; RIGHT: EMMANUEL GARNIER

High cirrus clouds over New York City. These clouds consist of small ice crystals and are at an altitude of 8 to 10 kilometers. © JOSHUA WOLFE

representative of a regional climate change (as discussed in Chapter 1). On the other hand, if a weather station is moved from the center of a city to a more rural airport site, this could introduce a spurious cooling jump into the record. For this data to be useful for climate monitoring purposes, then, climatologists need to adjust for these kinds of effects. These adjustments are determined through painstaking analysis of the written records, statistical tests, and correlations with other nearby stations.

Meteorologists also are trying to get to the heart of what goes on higher up in the atmosphere. The most complex processes take place at the level of individual cloud droplets and the aerosols that drift around the atmosphere. These processes determine when clouds form, how long they last, and whether they rain, thereby controlling the net effects of clouds on the climate system itself. A key uncertainty is the life cycle of the ice particles that make up high-level cirrus clouds, since they have both a reflective (cooling) and a greenhouse (warming) role. Knowing how they might change is a big determinant in whether cloud feedbacks will amplify or damp down climate change.

NASA's high-altitude research plane, the WB-57, is one of the only ways to get data directly from the higher levels of the atmosphere (above 10 kilometers [6 miles] or so), including the lower stratosphere where many key processes occur. The plane carries no passengers, so instruments need to be designed to be extremely hardy (temperatures can drop to –80°C [–112°F]) and work independently of any operator. The plane is often used in conjunction with measurements from the ground and from balloon-borne instruments to pin down what is going on in a particular air column. These data are particularly useful for testing theories about radiation transfer and convection processes.

Ice is important for clouds, but it is also the overwhelming focus of scientists working in the high Arctic. During the Surface Heat Budget of the Arctic (SHEBA) project described at the beginning of this chapter, hundreds of oceanographers and polar specialists worked on dozens of measurements on the floating base camp. There, scientists measured the heat fluxes through the ice, the effect of melt ponds on reflectivity, and the interactions of oceans and clouds—all key parameters influencing Arctic and global climate. Among the analyses made at SHEBA are fundamental studies on how the reflectivity of the snow and ice changes as a function of temperature and surface conditions, and how clouds and sea ice interact to control the heat balance of this crucial component of the climate system. None of this necessary knowledge could have been determined without individuals going to the Arctic to make the measurements themselves.

The perils of living on an ice floe. Constant movements of the ice, which is driven by the winds, occasionally cause the ice to ridge, with one ice floe crunching into another. Any equipment caught in between can be crushed or lost. © BRUNO TREMBLAY

In both field experiments mentioned above, there was an unseen partner: the network of orbiting satellites that crisscross the sky. More than one thousand functioning satellites are currently in orbit—some are military, some are for global-positioning systems. But many are Earth Observing Systems, put there to measure the intensity of the light across the whole electromagnetic spectrum; to recover information on clouds, water vapor, temperature, ozone, and soil moisture; and to measure the amount of particles in the atmosphere. They probe variations in the ocean and ice surfaces with laser altimeters, and measure tiny variations of gravity due to seasonal and long-term changes in water storage. The great advantage of satellites is the global (or near-global) view they can provide. But satellites also have limitations: cloud-viewing satellites, for instance, can not see low-level clouds if they are underneath higher clouds, and satellites looking at aerosols often have problems picking them out if the land beneath is not very uniform. Regardless of what they see, satellites send out a large continuous stream of data back to Earth. On the receiving end, researchers face the problems of data overload and spend most of their time trying to reduce the terabytes of incoming data into something more manageable. Sometimes a satellite's instruments suffer slow changes in accuracy that can be difficult to detect, although if there is an overlap with another similar satellite, they can be recalibrated. Over time, as well, the satellites gradually slow and their orbits decay due to very slight friction with the upper atmosphere. This degradation also needs to be detected and adjusted for.

Because of their near-global reach, plans are continually being made to upgrade satellite instruments and replace aging hardware. This process is very long and expensive; it can take a decade for a new measuring device to actually make it into orbit, and there have occasionally been gaps when instruments failed before their replacement could be launched. These service gaps, and the constant need for adjustments to correct for known problems, mean that long-term satellite-derived trends are, at present, a supplement to, and not a complete replacement for, ground-based observations.

However, the top-down perspective of satellites is very useful for the large-scale context, and is in sharp contrast to the necessarily limited viewpoints of ground observers. Since the satellite era began in earnest in 1979, the completeness of the observations has increased markedly. This bonanza is most noticeable in weather forecasts, which improved dramatically in accuracy when the satellite data began to be used. For climate models, as well, the satellite data on solar activity, precipitation patterns, radiation fluxes at the top of the atmosphere, and cloud variability have all added unique ways to test and improve these simulations.

RECONSTRUCTING PAST CLIMATE

The amounts of data being generated by the current observational networks and satellites are truly phenomenal, though still not sufficient for many purposes. They are incapable of providing a history of climate change before the beginning of the satellite era, or before the implementation of the weather station network in the late nineteenth century. Without such a longer-term overview, it is impossible to place current climate change in context or to understand what the climate is capable of doing. Information about past climates is buried in a multitude of archives, some well known (tree rings, pollen) and some obscure (isotopes in stalagmites, pack rat middens, chemical biomarkers in ocean mud). Each can be analyzed to say something about temperature, rainfall, or ocean or atmospheric circulation in different periods of the past.

The tree rings and corals mentioned in Chapter 1 have very clear annual bands, making them easy to date. Plenty of other archives exist that lack such precision but are still invaluable for detecting climate changes, both recently and further back in the past. One of the most promising is the study of cave deposits, usually stalagmites, which can be well dated using the decay series of uranium in the limestone.

Jeff Dorale of the University of Iowa holds a stalagmite from Crevice Cave near Perryville, Missouri. Stalagmites hold a record of climate change in their layers, which build up over time. Here, the stalagmite has been sliced in half, revealing its layers. The record of oxygen isotopes in each layer tells us something about temperature and rainfall. The large stalagmite in the background is about 130,000 years old. © PETER ESSICK

Stalagmites form as groundwater rich in dissolved minerals drips into a cave, where the minerals precipitate. They can grow quickly or very slowly, depending on location, and individual stalagmites can be decades to thousands of years old. Careful geochemical analyses of the limestone (calcium carbonate) layers can provide continuous, well-dated records of climate change in the cave region spanning millennia into the past. One of the most useful measurements is the ratio of oxygen isotopes in the carbonate, which derives from the isotope ratio in the water dripping into the cave.

Oxygen atoms come in a number of stable isotopes that differ in the number of particles in the nucleus. Chemically, the oxygen behaves the same way regardless of whether the nucleus holds sixteen, seventeen, or eighteen protons and neutrons. But some physical processes differentiate between the heavier atoms and the lighter ones. For instance, during the evaporation of water (H_2O), water containing the lighter atoms (written ^{16}O) evaporates more readily compared with the heavier ones (^{17}O and ^{18}O). When a cloud condenses, heavier atoms form water droplets more readily than lighter ones. The isotopes in the rainfall contain heavier oxygen isotopes at first (a greater proportion of ^{17}O or ^{18}O), and then progressively lighter ones (a greater proportion of ^{16}O) as more of the water vapor rains out. Changes in the isotopic ratio in the stalagmite over time can then be connected to changes in rainfall patterns and local temperatures. It's as if some oxygens were painted a different color. By looking at the eventual blend, we can work out how the proportions were mixed.

Because water is ubiquitous in our climate, many archives contain a record of water isotopes. Most directly, the snow and ice falling on mountain glaciers and ice sheets carry with them information about the source and temperature of the water. As the snow accumulates, layers of ice form, each with their own isotopic signature, laying down a history of change. In the polar regions, the dominant effect is related to temperature; as it gets colder, there are fewer heavy atoms (because the colder air has lost more of its initial water vapor). In the tropical ice cores, the signal is more tightly tied to patterns of rainfall change.

The deeper you go into the ice, the older it will be. In the polar ice sheets, we find ice that is more than 100,000 years old in Greenland and at least 850,000 years old in the center of Antarctica. The pioneer cores were drilled in Greenland in the 1980s. But it wasn't until the 1990s that two cores more than 3,000 meters (9,800 feet) deep extracted at the Summit site (in the middle of Greenland) really demonstrated the value of these archives. Drilled separately by U.S. and European teams 30 kilometers apart over a period of years, these cores were remarkably similar and told a story of large and dramatic changes. Particularly for the last ice age, these findings overturned widely held notions about the stability of past climates. Since the 1970s, conventional wisdom in the science community was that ancient climate change had been slow and driven predominantly by changes in the Earth's orbit. Measurements of water isotopes, dust accumulation, and gases such as methane (trapped in air bubbles in the ice) demonstrated that temperature swings of tens of degrees could happen in only decades, with severe consequences for the climate across the hemisphere. The last time such a swift change occurred was around 12,000 years ago.

The most dramatic result from the study of ice cores is the strong connection between Antarctic temperatures and the concentration of carbon dioxide in ice core bubbles. Carbon dioxide is not well preserved in Greenland cores because of the

Sylvia England processes an ice core from the West Antarctic Ice Sheet at the National Ice Core Lab in Denver, Colorado. The core is sliced and diced before being sent to different labs for analysis of the gas trapped in the bubbles, and the isotopes and dust in the ice. © JOSHUA WOLFE

high and variable dust content that is a consequence of the number of deserts in the Northern Hemisphere and Greenland's relatively southern location. But in Antarctica the snow is significantly cleaner. Scientists have managed to extract very accurate records of greenhouse gas concentrations through at least 800,000 years. Over eight glacial cycles driven by wobbles in the Earth's orbit, the temperatures and greenhouse gases moved up and down in lockstep, with the carbon dioxide occasionally lagging slightly behind. Here we have prima facie evidence that climate affects the carbon cycle (as discussed in Chapter 6), but it also hints very strongly that greenhouse gas changes play a role in how cold the ice ages are. Calculations suggest that, in fact, decreases in carbon dioxide, methane, and nitrous oxide provided just under half of the climate forcing that kept the ice ages as frigid as they were. The other factors were the extent of the ice sheets, the increased amounts of dust in the atmosphere, and the decreased amount of vegetation, which collectively reflected more solar energy to space. The cycling and slight lag indicate that, in these circumstances, the greenhouse gases acted as feedbacks on the orbital forcing and then contributed to the temperature changes themselves. More fundamentally, this cycling implies that, to a large extent, climate is predictable, and given a similar configuration of external factors, the climate would return to a similar state.

Oxygen isotopes come up again in the layers of muddy sediment that accumulate in oceans and lakes the world over. This mud is made up of the slowly accu-

A group of summer students from Lamont-Doherty Earth Observatory at Columbia University with scientist Tim Kenna taking a sediment core from the Manitou Marsh north of New York City. The mud they recover will be analyzed for pollen and charcoal—signatures of climate, human occupation, and the history of local fires. © JOSHUA WOLFE

mulating "snow" of organic matter that falls from the productive surface layers, building up over thousands to millions of years. Sediment records are retrieved by specially equipped drill ships that can drill down hundreds of meters into the sediment bed, in some cases in water depths of more than 4 kilometers (2 miles) and in some very rough seas. Paleoceanographers take a small sample of the mud and pick through it using a microscope to find the calcium carbonate shells of specific species of small zooplankton called foraminifera. Just as with the stalagmites, the carbonate oxygen isotope ratio gives us clues about past climate conditions—in this case, the local temperature and water mass properties. In ideal cases, the bottom of a lake or isolated ocean basin is anoxic—that is, not enough mixing of surface water down to the depths takes place to provide oxygen. Thus there is no marine life to disturb the bottom, and the sediment can accumulate in neat layers called lamina. These records have shown that the variability in the Greenland ice core is mirrored as far afield as the coast of California near Santa Barbara and in the Cariaco basin near Venezuela.

At different timescales, the ocean records can be a measure of ice ages waxing and waning or of oscillations of ocean circulation. Just as for ice cores, many other relevant parameters can be measured, each of which gives another partial view of the climate in centuries and millennia past. For example, the accumulation of wind-blown dust in some records off West Africa has offered a clear view of the rise and fall of the "Green Sahara." As depicted in the movie *The English Patient*, many areas in what is now the Saharan desert used to have abundant water and wildlife, including hippos and crocodiles. Now these areas are sandy desert, and the only obvious signs of these animals are in ancient petroglyphs etched into the rock. Dust from these dry lake beds is often blown out to sea by the winds, where it ends up accumulating in the ocean sediment. During the wet period following the end of the last ice age, and for many thousands of years, the lakes were full and the dust levels were much lower. However, dust levels found in Atlantic sea-floor sediment cores shot up very quickly for the period around 5,500 years ago, indicating widespread drying and desertification. To this day, the rains have never returned. We discuss the possible reasons for that below.

Mud from lakes, bogs, and the ocean contains a wealth of information about changes in plant life, as well. Climate scientists use these plant histories to validate the more indirect measures of climate change mentioned above. Pollen from each plant species is both small and unique. It is often airborne and, therefore, part of the dust that settles in lakes and oceans. From our research on plant communities today (see Chapter 5), we know which plant species are more heat- or drought-tolerant. Mapping the presence or absence of particular species at a given location over time yields a reasonably accurate portrait of the changes in the ecosystems there. In the Sudan, palynologists (those who study pollen) found that records taken from bogs and old lakebeds also show a switch from tropical savanna to the desert conditions of today at the same time indicated by the dust records. Large increases in ragweed pollen coincide with the arrival of Europeans and their woodland clearances on the East Coast of the United States in the seventeenth century. The expansion and contraction of the range of the alpine flower *Dryas octopetala* shows the coming and eventual going of the great European ice sheets. Pollen histories are often not as quantitatively specific as isotope records (the connection between exact climate conditions and pollen distributions are a little fuzzy at times), and the layers are often less easy to date than ice cores. Yet pollen histories can reveal many subtle features about how ecosystems and the local environments respond to climate and other disturbances.

There are few limits on how far certain scientists will go to find clues about past climate. Some will root around in piles of plant material stored in desert caves by pack rats. These small rodents solidify their nests into a midden with their highly viscous urine. Remarkably, in dry areas such as the Grand Canyon, these pack rat middens can last for tens of thousands of years and contain within them pollen evidence for the local plant life and climatic conditions. They have shown that the American Southwest was much more heavily wooded a thousand years ago. It is the combination of this paleovegetation data, the tree ring records of drought men-

Camille Holmgren holding a pack rat midden from a cave near San Simon, Arizona. The middens in the cave date from about 14,000 years ago, the transition period between the last ice age and the Holocene. Pollen from vegetation such as pinon pines and oaks were present at that time, indicating a wetter climate in the region than today.
© PETER ESSICK

tioned in Chapter 1, and the insights from local archeology that then allows the histories of the local cultures to be pieced together.

In the tenth century, the population started to rise in the Four Corners region of the Southwest as maize cultivation arrived from Mexico. The number and density of these Ancestral Pueblo settlements grew, as evidenced by the impressive multi-story cliff dwellings they left behind. However, near the peak of their cultural influence in the twelfth and early thirteenth centuries, many sites were abandoned. The coincidence in timing of this event with the deforestation, the megadroughts, and the establishment of new sites closer to the Rio Grande valley suggests a strong two-way interaction between human culture and the environment. The forests likely fell victim to a combination of human activity and periodic droughts, and the ability to sustain such population densities likely relied on the forests themselves, as well as a regular water supply.

TRACKING ECOSYSTEM IMPACTS

Obvious signs of climate change include a shift in the physical environment from one ecosystem to another. But careful measurements and clear understanding are needed to be able to link the ecological changes, such as plant disappearances, to the physical climate, such as the temperature and rainfall. This work can be extremely painstaking. Such detail is critical, however, and includes cataloging and quantifying biodiversity in various ecological niches and tracking how this changes as a function of year-to-year variability and eventually long-term trends (as described in Chapter 5).

It appears to be a fact of nature that the most sensitive and fast-changing components of the ecosystem are the smallest and hardest to observe, such as insects, plankton, and parasites. However, these are often key players in controlling the impacts of change on the larger-scale species that we are more normally interested in, for example, the impact of krill abundance on penguin colonies (see Chapter 5). Therefore, long-term ecosystem monitoring must be very detailed if it is to be useful. In areas where careful long-term studies have been done (such as in the cloud forests of Costa Rica or the dry valleys in Antarctica), changes due to local and large-scale climate change have been profound.

It is often only in hindsight that the value of long-term records becomes clear, and it is remarkable that many of these records are the result of the perseverance and painstaking work of only a few determined individuals. In any specific ecosystem, the impact of climate change will be very individual and thus difficult to generalize to the global scale. Since many ecosystems are not being monitored as closely as they could be, many changes are likely going unrecorded.

Researchers measure the carbon storage in soils near the Monteverde Preserve in Costa Rica, in order to help determine the fate of the human emissions of fossil fuel carbon. © GARY BRAASCH

MODELERS

There are multiple interlocking themes, such as the basic cycles of energy and carbon, that link the different communities of climate researchers. For instance, the water cycle links the water isotopes measured in the ice cores to the drought reconstructions from tree rings, to the ice measurements at SHEBA, to the cloud forest habitat of amphibians in Costa Rica. How are such links evaluated and quantified? This is where the climate modelers come in.

Some climate models are relatively simple, such as those that try to encapsulate the total energy budget of Earth's total inputs and outputs as discussed in Chapter 6. However, the most common models used to examine climate change on the decadal

In the early years of climate modeling, all instructions were input into a mainframe computer using punch cards. Each card was good for one line of code, and a whole program would require hundreds or thousands of cards, all in the exact right order. These cards were saved from an Amdahl V/6 machine used by NASA in the 1960s. © JOSHUA WOLFE

to centennial timescales are called coupled ocean-atmosphere General Circulation Models (GCMs). These GCMs are developed by approximately fifteen separate modeling groups from around the world. They contain much of the physics of weather, ocean circulation, and sea ice, all of which is tied together so that changes in one part of the simulated world have a multitude of effects on all the other components. Some newer models are even more complex. Earth system models (ESMs) include the physics of the GCMs, but also simulate the interactions between aerosols, atmospheric chemistry, vegetation, and ocean biology. The climate models are well known for the projections described in the Intergovernmental Panel on Climate Change reports (and in Chapter 8). These models involve huge numbers of processes that interact in complicated and nonlinear ways. This inherent complexity could make us uneasy if the results were presented as answers that appear to come from a black box whose inner workings are opaque. This is a valid concern, so it is worth uncovering how scientists develop confidence in the projections of such climate models.

Climate modelers have set themselves a formidable task—to take our knowledge of the equations of fluid motion for winds and ocean currents, the physics of radiation through the atmosphere, and estimates of the effects of clouds, convection, and the fluxes that connect the different physical components, and use that knowledge to understand the large-scale features of the climate system and their response to external pressures. All of these calculations must be done for a system whose key components are in constant motion. The idea to numerically predict weather was first formulated in the 1920s, but it wasn't until the 1960s that computers started to make the large-scale calculations feasible. The essential concept is that the atmosphere can be divided up into small chunks (called grid boxes) that are considered homogeneous and that interact with neighboring chunks through the winds, radiation, and other processes. The smaller the chunk, the better the approximation. Since computational power has grown exponentially, so has the detail that can be included in these models.

Weather forecast models include a lot of detail for atmospheric processes, but climate models have to include much more. Climate models need descriptions of land, oceans, sea ice, and more recently, interactive atmospheric aerosols, atmospheric chemistry, and representations of the carbon cycle.

We can divide the physics of climate that we need for models into three classes. First, we use aspects of fundamental physics, including the conservation of energy and mass, calculations of the Earth's orbit, and the distribution of the Sun's energy over the Earth's seasons. These basic components are calculated with as much precision as the computers allow. Second, we use physics that is very well known in theory, but must be approximated in some way due to time constraints or complexity. Examples include how much radiation goes through the atmosphere or the specific equations of motion for particles moving within the ocean or atmosphere. Some details get lost, such as the specific absorption of radiation at one particular frequency, or the impact of a small gust of wind, but the overall behavior of a system is well captured. Third, we use physics for which we only know empirically measured values. Such phenomena include the formula for calculating evaporation as a function of wind speed and humidity, which is derived from observations of the process in many different environments.

These empirical formulas are called parameterizations. The uncertainty in these formulas accounts for a lot of the differences between the models. However, much of the large-scale behavior seen in models is robust and does not depend on the details of the parameterization. Why? Because much of the interesting behavior of the climate system is *emergent*. For instance, we have no formula for modeling a midlatitude

Gary Russell, a climate model developer with NASA, examines old output from an early 1980s-era climate model. In the days before graphical displays, the only way to see model results was to read off the numbers. © JOSHUA WOLFE

Allegra LeGrande is typical of a new generation of climate scientists who are using the models to look at past climate changes, incorporating atmospheric chemistry and isotopes modules for more direct comparisons with the observed data. © JOSHUA WOLFE

storm track across the North Atlantic, yet all models show one. The storm track emerges instead out of the complex interaction between the warm tropics, the cold poles, and the certain properties of the equations of motion.

Why do scientists use such models? We do not use them just because they are complicated. Rather we use models because they have just enough complexity to help explain enormous amounts of what we see in the real world. The confidence we have in the climate models comes from tests they have successfully passed in predicting what is seen in the real world.

So how does that get done? The testing happens at two distinct levels. At the first level, individual processes are tested; each parameterization, such as the formula for evaporation, can be checked against observations. As the observations improve, the parameterizations can, too. At the second level, testing applies to large-scale emergent properties—the seasonal cycle in tropical rainfall, the statistics of winter storms, or the response of the atmosphere to an El Niño event. These tests are not directly evaluating any particular part of the model, but rather the combined effects of all of the components.

The most relevant tests for climate change purposes are those that use natural experiments such as the Mount Pinatubo eruption to test the sensitivity of the climate to those forcings. Many consequences of this eruption were observed: we already mentioned the subsequent global cooling of about 0.5°C (1°F) (in Chapter 6), but in addition there was warming in the stratosphere, a decrease in global water vapor, and an increase in westerly winter winds in the North Atlantic. Each of these changes also occurs in climate models when we add the observed change in volcanic aerosols as the forcing.

Another example is the drying of the Sahara five thousand years ago, as mentioned earlier. Theories suggest this was a natural phenomenon related to slow changes in the Earth's orbit. The key factor is that six thousand years ago summer in the Northern Hemisphere was warmer because the Earth was closest to the sun in July (rather than January, as it is at present). This made the tropical rain bands move farther north, making it easier for vegetation to grow. But complex vegetation

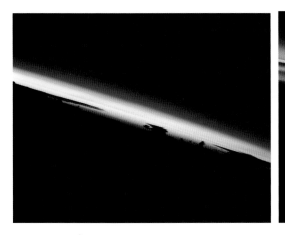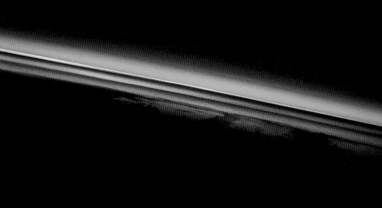

dynamics created a nonlinear trigger for rapid change (since there is a threshold of rainfall below which vegetation won't grow). Models do show this behavior as the Earth's orbit is changed.

The models are not perfect. For instance, the scale of key processes right on the equator is often smaller than the size of the ocean or atmosphere grid boxes used in the models. This makes it difficult to simulate El Niño events properly and gives rise to systematic biases in tropical rainfall, in that the models tend to have excess rain on either side of the equator, but not on the equator itself. Yet, despite those problems, the models give very consistent answers for how sensitive climate is to increasing greenhouse gases or to the spatial patterns of ocean temperature variations. Over the last twenty years, their success in matching observations has measurably improved, and as computer resources increase, this improvement is likely to continue.

The most important role for these models is their ability to quantify otherwise qualitative hypotheses and generate new ideas that can be tested against observations. They are extremely flexible tools, and although far from complete, the results so far have been surprisingly successful.

Observations of large-scale climate as described above are partial and sometimes apparently contradictory. Climate models can help assess whether this might be because of weather noise or whether a significant discrepancy really exists. In that case, both the models and the data need to be examined to see whether anything is wrong with the models, or as is sometimes the case, there is a problem with the data. This works well for resolving current climate conundrums with the amount of observations now available, but this should be just as useful for the larger changes seen in the paleoclimate record. There are additional problems in that case, because the interpretation of paleoclimate data is less clear than for mod-

These two photographs show the Earth's horizon at sunset before and after the 1991 Mount Pinatubo eruption. In the first picture, the atmosphere is relatively clean, and the colors change smoothly above the tops of the thunderstorm clouds. In the second picture, two months after the eruption, two dark layers of sulfate aerosols at 20 to 25 kilometers above the surface are clearly visible.
NASA

ern measurements. However, coincidentally, many of the new processes now being included in the Earth System Models, such as dust or the carbon cycle, are also the processes that produce the changes recorded in the ice cores or ocean sediment. In one particular example, some models now contain the physics of water isotope fractionation and can be compared directly to the isotopic archives mentioned above. These comparisons allow us to test hypotheses for climate change over a much larger set of examples than was possible even ten years ago, and hence improve our confidence in the models.

THE LINGUA FRANCA OF CLIMATE SCIENCE

The interconnectedness of the climate system and its rise as a coherent field of study is a dominant and integrating force in Earth sciences. Climate science brings together the researchers who drill in ice sheets with those who climb trees in tropical forests, those who study cloud microphysics with those who observe deep ocean circulations, those who study plant genomes with those who study rock formations. These experts in any other situation might never have had anything to do with each other.

However, we face numerous challenges in trying to communicate across disciplines. Specialized fields tend to have their own jargon, their own journals, and their own debates about what is or is not important. Often the specialists in a field will have developed a shorthand for describing a phenomenon—a description that may be at odds with how experts in a different field see that same phenomenon.

A good example of the challenges of multidisciplinary science and how they can be met is the effort underway to understand the variability and dynamics of the North Atlantic thermohaline circulation (discussed in Chapter 3). This effort to understand ocean currents links physical oceanographers with paleoclimatologists, modelers, and theorists who all have separate perspectives derived from the tools and techniques that they use.

Physical oceanographers collect very-high-resolution data with timescales of a few minutes to a few years. This data is mostly gathered from research cruises and fixed moorings (a tethered line of instruments that stretch from the sea floor to the surface). The important signals measured relate to ocean eddies, the seasonal cycle, and tides. Long-term changes over a few decades are measured in

Ocean mud accumulates year after year and can provide key information about past climate. This giant gravity corer was deployed by U.S. Coast Guard icebreaker *Healy* in the Bering Sea in June 2003 to collect up to 6 meters (20 feet) of ocean sediment. Depending on the location, this much sediment can provide 6,000 to 6 million years' worth of ocean climate history. PHOTO BY JAMES BRODA, WOODS HOLE OCEANOGRAPHIC INSTITUTION, 2003 (WWW.WHOI.EDU)

tenths or hundredths of a degree. Paleoclimatologists use archives that go back thousands of years to the last ice age or earlier. They can see much larger changes, sometimes tens of degrees over that longer time span. However, they have a much coarser view in time, with one data point every few decades at best, but more usually one data point per century or even longer. Meanwhile, modelers try to reconcile the microdata gathered by physical oceanographers with the macrodata culled by paleoclimatologists. But computational limits mean that they cannot model the details as finely as can be seen in present-day observations, nor can they simulate thousands of years of ocean evolution.

Initial mismatches of perspective limited the opportunities for collaboration for many years. However, available observations can now stretch back decades. Meanwhile, some paleoclimatologists have switched their focus to the recent Holocene period (the last 10,000 years), and some very-high-resolution ocean sediment data (down to individual years) is starting to become available. Finally, as computational resources have improved, the simulations of ocean variability now overlap with both direct observations and paleoclimate data. As a result, each of the groups has started to see merit in the others' techniques and conclusions (since these might now illuminate their own studies), and constructive engagement on theories of ocean change has become easier to achieve.

Consider the following example. About eight thousand years ago, the North Atlantic region experienced an episode of abrupt cooling. The leading hypothesis is that the rapid drainage of what had been a large ice-dammed lake into the Hudson Bay reduced the density of the North Atlantic Ocean and slowed the heat transport of the thermohaline circulation (see Chapter 3). The old lakeshore of Lake Agassiz, a precursor of the Great Lakes we know today, can still be seen in Ontario and Manitoba and can be dated to the same time. To see if this idea is correct, paleoceanographers have analyzed new ocean sediment data to provide evidence of ocean changes at the time. Modelers are trying to simulate the observed changes by adding the freshwater from the lake into their modeled North Atlantic to see what effect it has on the ocean density and circulation. The physical oceanographers are using their experience with the complexity of the modern ocean to interpret the paleoclimate evidence and model results. The outcome has been a validation for both the idea and the models. The amount of freshwater that was released when the lake drained was apparently sufficient to affect the ocean circulation, and the temperature and rainfall changes that would have resulted match up with what is inferred from the proxy data. Since this episode is the most recent example of a significant shift in the North Atlantic circulation, it has a pivotal role to play in evaluating the importance of ocean circulation changes on climate. Effective collaborations on

A self-portrait by Ken Mankoff, a researcher working on ocean core drilling in McMurdo Sound, while taking a break at the official South Pole monument in late 2007.
© KEN MANKOFF

this topic thus lead directly to effective assessments of potential ocean circulation changes in the future.

More generally, the development of Earth System Models is forcing modeling centers to reach out beyond their core expertise and bring in people who are knowledgeable, for instance, about vegetation responses to climate. This both enhances the flexibility of the new modeling tools to encompass a wider set of scientific questions, and educates a new community about the usefulness of modeling at this scale. Possibly the most important new set of collaborations will focus on the role of ice sheets and their response to climate. Current GCMs assume that the response times for ice sheets are centuries or longer and so, historically, they have not included a dynamic ice sheet component. Recent observations, however, indicate that ice sheet changes can happen much faster than previously supposed (see Chapter 2), and the uncertainty in this response is dominant in discussions of future sea level rise (see Chapter 8). Multiple efforts are ongoing to improve this situation, and many of the same disciplinary challenges (of scale, language, and outlook) discussed above are apparent in this case as well. Given the history of climate science, it is likely that these challenges will be met, allowing for a much deeper understanding of past, present, and future ice sheet development. To answer the questions posed at the beginning of this chapter, scientists put themselves to so much trouble because they know how interconnected the climate system is. They also know that the answers we need can often only be sought in inaccessible, cold, dirty, and difficult environments. Perhaps surprisingly, they appear to enjoy meeting these challenges.

Peter Essick

SCIENTISTS STUDYING CLIMATE CHANGE

Scientists search for clues about the climate system in many places, some obvious, some obscure. Peter Essick searches for scientists.

Finding useful natural archives of past climate change is challenging because the conditions necessary to preserve these records are very rare. Greg Wiles, for instance, is analyzing the tree rings in trees that have been uncovered by the retreat of the Columbia Glacier in Alaska. These trees were knocked down and covered in ice around a thousand years ago. As the ice melts around them, they provide information about the timing and extent of past glacier changes. By contrast, we see drillers on the Quelccaya Ice Cap in Peru retrieving an ice core before the retreat of the ice erases this record forever.

Sometimes, though, you just need to dig. Jeff Pagati is sampling the distinct *Coro marl* (the white layer) at Murray Springs, Arizona, deposited during the last ice age (from 40,000 to 15,000 years ago), when local conditions were much wetter than today. The types of fossils found within the sediment help researchers determine local climate changes and impacts of the past.

The ice age may be long gone, but every winter Katie Hein and Dave Harring measure ice thickness on Lake Mendota, Wisconsin, adding new entries to an invaluable record that goes back to 1852. Monitoring current climate year by year is essential for detecting long-term, subtle changes, and tracking the ecosystems that are affected by those changes is a painstaking, and sometimes lonely, undertaking. Penguin researcher Bill Fraser, on Litchfield Island off the west side of the Antarctic Peninsula, has been keeping track of Adélie penguins for thirty years. From one thousand breeding pairs here when he started his research, there are now only twenty. Locally, temperatures have increased 3°C over this period and sea ice has declined significantly, reducing the habitat for krill, the main food source for these penguins. Ecologists aren't just paying attention to the charismatic fauna, though. High-altitude flora such as the lowly bryophytes (mosses) being counted by Daniela

Hohenwallner on Mount Schrankogl, Austria, can be sensitive indicators for alpine change.

Data can come from many sources, and translating the digital stream from Earth-orbiting satellites into arresting imagery is the job of scientists like Jason Sun at the Visualization Lab at the University of Texas, Austin. Gathering the different sources of information and revealing a few of Nature's secrets is, however, a rare talent. For fifty years, Wallace S. Broecker, the subject of the final, panoramic photograph, has been a dominant figure in the study of past climate, and he has played a key role in broadening our understanding of the ice age cycles and the nature of abrupt climate change. Those insights have led him to warn that humanity should be extremely wary of "prodding the climate beast."

Greg Wiles
examining
tree rings near
the Columbia
Glacier, Alaska

Victor Zagorodnov and Vladimir Mikhalenko handling a recently drilled ice core from the Quelccaya Ice Cap, Peru

Jeff Pagati sampling the *Coro marl* (the white layer) at Murray Springs, Arizona

Katie Hein and Dave Harring measuring ice thickness on Lake Mendota, Wisconsin

Bill Fraser studying Adélie penguins on Litchfield Island, off the west side of the Antarctic Peninsula

Daniela Hohenwallner counting bryophytes (mosses) and other high-altitude flora, Mount Schrankogl, Austria

Jason Sun, researcher in the Visualization Lab at the University of Texas, Austin

Wallace S. Broecker at the Lamont-Doherty Earth Observatory

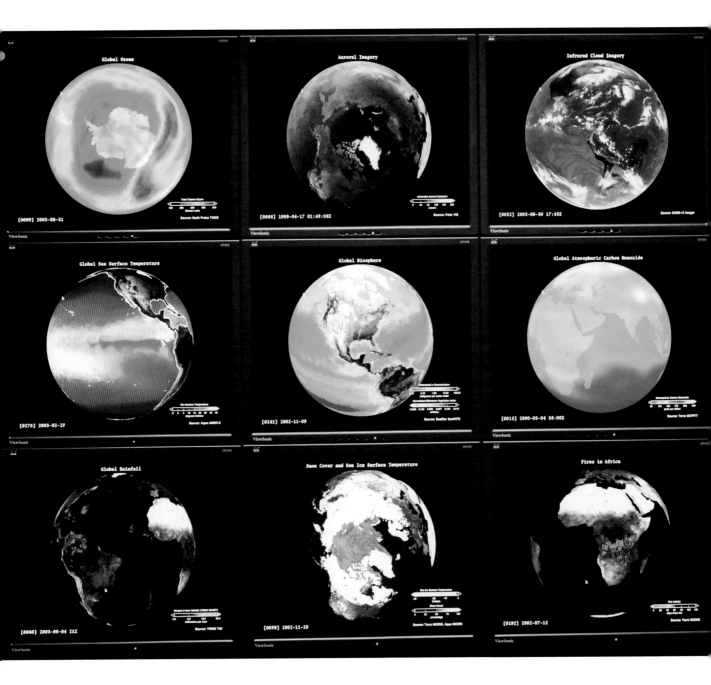

CHAPTER 8

THE PROGNOSIS FOR THE CLIMATE

Gavin Schmidt

It is very difficult to make an accurate prediction, especially about the future.

> **—Niels Bohr, 1922 Nobel Laureate in Physics,
> possibly paraphrasing Mark Twain**

The future ain't what it used to be.

> **—Yogi Berra**

So what is to come? Fortune-telling is beyond the ability of scientists, (and in any case would be a class B misdemeanor under the New York State Penal code section 165.35), yet we have a clear need to explain the implications of our scientific understanding for future climate.

FORECASTS, PREDICTIONS, AND PROJECTIONS

It is in the prognosis for what the future may hold that the medical analogy that runs through this book is most apt. Just as doctors use their medical training to give a preliminary diagnosis for a patient, climate scientists do the same for the planet. Just as doctors order more tests to confirm (or refute) that diagnosis, climate scientists use information about past climate changes to refine our understanding of what's happening now. Finally, just as doctors try to give a prognosis for the future course of a disease or illness, climate scientists are in a position to say what course climate change may take. These are not definitive statements about what must hap-

Keeping tabs on climate with the Hyperwall at the NASA Goddard Space Flight Center. © JOSHUA WOLFE

Global Mean Temperature Projections

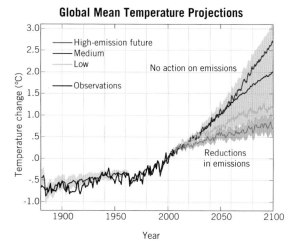

The global mean temperature projections for a number of scenarios using the NASA/Goddard Institute for Space Studies climate model. The twentieth-century changes are in black; the model's simulation for that period are a good match to the observations, particularly after 1940. For future projections, three business-as-usual scenarios are hypothesized, along with one projection based on a strong effort to mitigate emissions. The important points to note are that even the scenario with mitigation shows substantial future warming (another 0.7°C above 2000 levels) and that the temperature changes in the other scenarios are up to four times greater than the changes seen in the twentieth century. Other models in the IPCC simulations show even more dramatic changes. Choices that we make as a society will determine which path we follow. GAVIN SCHMIDT

pen, but more cautionary conclusions about what might happen if current trends continue. Thus, a climate projection of increased warming is analogous to a doctor's warning that if you continue to put on weight, you will be at increased risk of heart disease.

Much of the confusion in the popular press about climate projections stems from the ambiguity of some of the language used, and in particular, the differences between *forecasts*, *predictions*, and *projections*. Colloquially, these three words are interchangeable. Although the scientific community has tried to give them very specific definitions, those definitions are not universally adhered to; sometimes not even by the scientists themselves. However, the distinctions are useful and important.

What are the "official" definitions? A *forecast* is a description (sometimes given in terms of probability) of what is going to happen in the future. For example, today's weather forecast could include a 60 percent chance of rain this afternoon and a peak temperature of 25°C (77°F). This description contains some uncertainty, but it is well accepted as a good estimate of what will actually happen. A scientific *prediction* is a much more general concept, a statement that under a particular and precise set of circumstances, something will happen. For instance, if there is an El Niño event, California can expect a higher probability of rain. A prediction in this case is not exclusively tied to statements about the future, but can apply to predictions of what might be discovered about some past event, or be seen in present-day observations. These "if, then" predictions are the fundamental building blocks of the scientific method, because they allow for the potential refutation of the underlying theory or hypothesis. If under the same set of circumstances, Y happens instead of X, then the original theory is proved at best incomplete or at worst completely wrong. Note that the set of circumstances in question has to be very tightly controlled for this to work. Too many uncontrolled variables make it very difficult to draw conclusions. Controlling variables can be easy for scientists studying small systems in the laboratory, but it is more difficult to arrange in observational sciences such as climatology, in which the laboratory is the whole climate system, or epidemiology, in which it's the entire population. In such cases, cause and effect are harder, though not impossible, to determine.

Finally, we have *projections*. If the models we are using have proved useful in simulating a wide range of situations in the past, then we can use them to suggest what might happen in the future. Think of projections as conditional "what if" forecasts or as a special class of predictions. In the climate context, we can project a particular kind of climate change *if* emissions continue to rise at a certain rate (defined by an emissions scenario). Note that this statement is different from a forecast since no claim is made that the particular scenario will ever come to pass. Implicitly, we assume in making a projection that all else is equal—in other words, factors other than emission increases are not also affecting the climate. The scenarios that the Intergovernmental Panel on Climate Change (IPCC) uses for its projections (discussed in more detail below) are significantly more uncertain than the climate science used to estimate their consequences. The uncertainties in each step add up, which is the principal reason climate forecasts (in the precise sense of the word) are impossible over the long term.

The IPCC projections are not forecasts for another important reason. Whatever happens to global temperature in years to come, it will be a function of two things: the forced or driven response, which is to a large extent predictable, and the unforced weather component, which is contingent upon the actual path of storms, the ocean currents, or the randomly timed eruptions of large volcanoes. Atmospheric weather conditions have what is called a "sensitive dependence on initial conditions" and are chaotic (in the technical sense of the word). This means that tiny differences in how you start a weather forecast lead to dramatically different results after two weeks or so, which is the reason weather forecasts are only accurate for a short period. Even theoretically, they are never going to be better than a week or so. The ocean has its own kinds of "weather"—the turbulent changes in Gulf Stream eddies or El Niño in the Pacific—that changes more slowly than the weather in the atmosphere. Ocean conditions, too, are thought to be chaotic, but because they have more persistence, we have some evidence they are useful for seasonal (and maybe multiyear) forecasts.

True forecasting, either for weather or for ocean circulation, is an initial-value problem. That is, if we know the weather conditions today, we can use our knowledge of fluid mechanics and meteorology to forecast how weather conditions will develop in the future. The IPCC projections are instead boundary-value problems. They do not depend on today's weather. Rather they respond only to the changes in external drivers such as greenhouse gas amounts. In the future, true climate forecasts may in fact be possible. We are starting to develop the capacity (through the Argo float network mentioned in Chapter 1) to collect enough information about the ocean's exact state that we may soon be able to start off our climate models with an observed initial state of the ocean. If the models are good enough, we may be able to predict short-term (up to a few years or so) changes in the ocean, which will then give some predictability to the atmospheric weather statistics. The price we pay for not being able to do this modeling at the moment is that today's projections are only good for average changes over the long term and not for next season or next year.

THE LOGIC BEHIND CLIMATE SCENARIOS

The first requirement for a climate model projection is to describe the change in external conditions. This change can be idealized, such as instantly doubling the level of carbon dioxide (CO_2) in the atmosphere (a so-called $2xCO_2$ simulation) or increasing CO_2 levels by 1 percent a year, and these tests are useful for comparing results across models. Much more complicated and realistic emissions scenarios, however, are used by the IPCC.

Social demographers and economists develop the story lines for the IPCC scenarios. These story lines consist of projections for population, technology changes, economic development, resource limitations, and the like for the next few decades to a century. If we consider how much energy people are likely to use, what technology will produce that energy, and combine this information with world population projections, we have an estimate of what emissions will look like in the future. A huge range of issues are built into these scenarios: how quickly the world economy will grow, how much developing countries will progress, what technological breakthroughs will define the twenty-first century, and so on. Rather than explore every single hypothetical pathway, climate modelers generally use high-, middle-, and low-marker scenarios. They use those markers to bracket plausible changes, but because many of these uncertainties are unknowable, they do not generally make a statement about which scenarios are more likely. The widely used IPCC scenarios assume that we will not reduce emissions because of climate change. So, these IPCC projections are collectively described as business-as-usual scenarios.

Although these scenarios do contain many of the factors that scientists think will be important in the future, they were not complete. Factors such as changes in ozone precursors were not defined. The impacts of climate on the carbon cycle itself were not generally included, either. Neither were volcanic eruptions (which are unpredictable, of course), nor further long-term changes in solar activity (which are unpredictable as well, at least with our current understanding of solar physics).

Other scenarios have been developed that do account for efforts to control emissions. These include stabilization scenarios, in which it is hypothesized that CO_2 concentrations can be stabilized at, say, 450 or 550 parts per million, and what are sometimes referred to as alternative scenarios, in which it is assumed that aggressive measures begin immediately. For example, the "with mitigation" scenario in the illustration is one that assumes a 50 percent decrease in CO_2 emissions by 2050, and a 70 percent decrease in CO_2 and a 45 percent decrease in anthropogenic methane emissions, by 2100.

The plethora of available scenarios can be quite confusing, and the results from one particular scenario often are reported in the media as though they were certain predictions. A useful fact to note when interpreting such results is that climate impacts under any particular scenario are closely tied to the temperature change. What differs between the scenarios is when that temperature is reached.

HOW TO READ THE RESULTS

The projected climate responses to different scenarios are often given in ranges. For instance, if we follow a business-as-usual approach and don't adjust behavior to reduce greenhouse gas emissions, the temperature projection for 2100 goes from 1.1°C to 6.4°C (2°F to 11°F). It's important to know what these ranges mean. They generally encompass two sorts of uncertainty. The first (and most important, especially as you go further out in time) is the uncertainty due to the scenario itself. In the case of the temperature projections, for instance, the range of best estimates across the different business-as-usual scenarios is 1.8°C to 4.0°C (3°F to 7°F). The second source of uncertainty is seen when different models are given the same input. As discussed in Chapter 7, there are about fifteen separate modeling groups around the world, who all have made slightly different, but reasonable, decisions about how to model the climate. The range of responses in these models to the same input is a measure of our uncertainty in the science. The range of temperature for one particular midrange scenario across the different models is 1.4°C to 3.8°C (3°F to 6°F). The range usually seen in the headlines is the minimum value for the scenario with the lowest emissions (1.1°C) to the maximum value for the scenario (6.4°C). Thus, much of the uncertainty is due to the inherently unpredictable scenario, not our understanding of climate.

The most useful interpretation of these ranges depends on the purpose, but in general, the high end corresponds to the smallest number of technological improvements and the unluckiest breaks in terms of how sensitive the climate is. Similarly, the low-end estimates will come to pass only if we are technologically lucky and the models are overestimating climate sensitivity. The midrange values, although they can't be described formally as the most likely, are what you will generally get using middle-of-the-road assumptions about the future and the canonical value of the climate sensitivity. The worst and best cases cannot simply be dismissed as outliers. Sensible policymaking needs to factor in those possibilities as well. In this chapter, we give the best guess and the range, making clear when the uncertainty is due to the scenarios or to climate model uncertainty.

Regional temperature projections from IPCC show a little more information. For the period to 2030, the trends are already set and do not depend much on the scenario. Warmer temperatures are seen over Northern Hemisphere continents and in the Arctic—very similar to the patterns seen to date. For the situation at the end of the twenty-first century, the different emission scenarios make a big difference. In the low-emissions case, temperature changes reach 3°C over the continents and 4°C or 5°C over the Arctic, and more than a degree Celsius of warming is seen over the tropical oceans. However, in the high-emissions case, the changes are much worse—5°C to 6°C over the continents, 3°C in the tropical oceans, 3°C to 4°C over Antarctica, and even more over Greenland. These temperatures would take the ice sheet regions into temperature regimes they haven't experienced in millions of years, with unknowable consequences for sea level rise. MODIFIED FROM THE INTERGOVERNMENTAL PANEL ON CLIMATE CHANGE

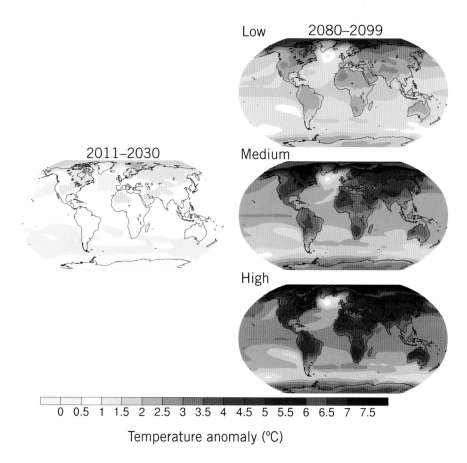

Temperature anomaly (°C)

We use two points in time to guide our discussions—the year 2030 and the year 2100. Most of us will still be alive in 2030. This earlier time is useful because it is close enough to the present to be clearly relevant to today's population. It is also near enough that the uncertainty in scenarios is minimized. According to projections, substantial differences in response between scenarios do not show up until around 2030, but by 2100 those differences become very significant and the impact of policy actions most noticeable. This should not be taken as implying that climate change doesn't continue after 2100—it certainly will!—but detailed projections out that far become very problematic.

RISING SURFACE TEMPERATURES

It is not surprising that the most commonly cited projection for future global warming is the likely rise in global average surface temperatures. Regardless of which projection or model we use, all the results indicate a little less than 1°C (2°F) additional warming from 2000 to 2030. Surface temperature changes will be about 50 percent higher over land than over the ocean and higher in the Northern Hemisphere than in the Southern.

By 2100, global average temperature changes will be much more significant. In a midrange scenario, the annual land temperatures will be 3°C to 4°C (5°F to 7°F) warmer, even though the global mean change is "only" 2.4°C (4°F). For the high-end projections, land temperature changes reach 5°C or 6°C (9°F to 11°F) higher than in 2000 in the higher northern latitudes (4°C or 7°F in the global mean). To put this into context, the global mean temperature during the peak of the last ice age 20,000 years ago was about 5°C (9°F) cooler, a level associated with what can only be described as a different planet. Compared to the estimates of temperature variations over the last one thousand years, the projected changes are five to ten times larger.

We can compare these projections with the changes that occur as a function of natural variability from season to season or year to year. For instance, over the last thirty years in New York City, the difference between the very coolest and the very warmest year is around 2.5°C (4.5°F). So a change of a few degrees in mean temperature implies that what are considered very warm years today will become the norm, and the new warm years will be unlike anything seen in the observational record. Warm temperature extremes, such as the number of days above 32°C (90°F), will also increase as the mean temperature does. Conversely, the number of cold winter nights will decrease.

As has already been described in Chapter 2, the Arctic regions will likely have the largest temperature trends. Model results indicate an average increase in the temperature of the Arctic of about 3°C to 5°C (5°F to 9°F) over land and 7°C (13°F) over the ocean by the end of this century. The projection of greater warming over the polar ocean is principally due to the decrease in extent of Arctic sea ice. Some projections suggest that the Arctic may lose its summer ice cover toward the end of the century, while others suggest the possibility of more rapid changes, perhaps even a total loss of summer ice by as early as midcentury. Permafrost thawing and degradation is projected to increase, perhaps reducing its current area by 10 to 20 percent or more by 2100. Warming temperatures mean that the southern limit of permafrost will shift toward the north, and the area of discontinuous permafrost will increase. Because North American and European permafrost is warmer and thinner than permafrost in Siberia (where it is significantly colder), it is projected to suffer the greatest degradation in the twenty-first century.

RAINFALL CHANGES

Projected changes in rainfall *patterns* are generally robust, even if the projected local amounts are not. The models suggest that rainfall will increase at mid- to high latitudes and near the equator, and decrease in regions of the subtropics that are already dry. The wet areas will get wetter and the dry areas drier.

Annual Rainfall Change by 2100

Changes in annual rainfall by 2100, as projected by the IPCC models. Blues indicate more rainfall—seen in the high latitudes and on the equator—and yellows and reds indicate drier conditions. These are most pronounced in the subtropical zones, including the Mediterranean, the American Southwest, parts of Amazonia, Southern Africa, and Australia. The stippling denotes where a substantial majority of the models are in agreement about the direction of change. MODIFIED FROM THE INTERGOVERNMENTAL PANEL ON CLIMATE CHANGE

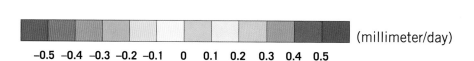

(millimeter/day)

–0.5 –0.4 –0.3 –0.2 –0.1   0   0.1 0.2 0.3 0.4 0.5

However, in some regions, such as the Sahel, climate models disagree strongly on what will happen. Some indicate that the Sahel will become much drier, others indicate it will become wetter, and still others predict little or no change. At present, we have no way of determining which is right, and we don't yet understand the reasons for the difference. Most of the models do quite a good job of simulating the Sahel drought in the 1970s when given the observed ocean conditions, so this uncertainty is potentially linked to differing projections of ocean circulation, particularly in the tropics (see Chapter 4).

In some areas, we have already observed that rain is falling in more intense bursts as temperatures increase and water vapor amounts rise. This trend is projected to continue. In the Arctic, the frequency of rain-on-snow events could increase by 40 percent, speeding the melt of snow-covered areas.

Changes in rainfall and temperature will combine to affect river flow and soil moisture. In mountainous areas, reductions in glacial extent will ultimately reduce the meltwater flowing into rivers in spring and summer. In continental interiors that are already very warm during the summer, reductions in meltwater and rainfall will lead to reductions in soil moisture and to even larger temperature changes.

RISING SEA LEVELS

Sea levels will rise for several reasons, some of which we can predict better than others. Principally, water expands as it warms (thermal expansion), and less ice on land means more water in the ocean. As the oceans continue to warm, increased sea level through thermal expansion is certain. This component of sea level rise is relatively predictable and will likely be around 25 centimeters (approximately 10 inches) out to 2100, but will continue for centuries as the very deepest parts of the ocean slowly warm. This is worrying enough, but uncertainty in the other components of sea level rise is even more threatening.

The impact of a 1-meter sea level rise would be seen all around the world, from Bangladesh to the Netherlands to the Eastern Seaboard of the United States. It's estimated that over 60 million people would need to be relocated, and many millions more would be significantly more vulnerable to storm surges and coastal flooding.
© NASA

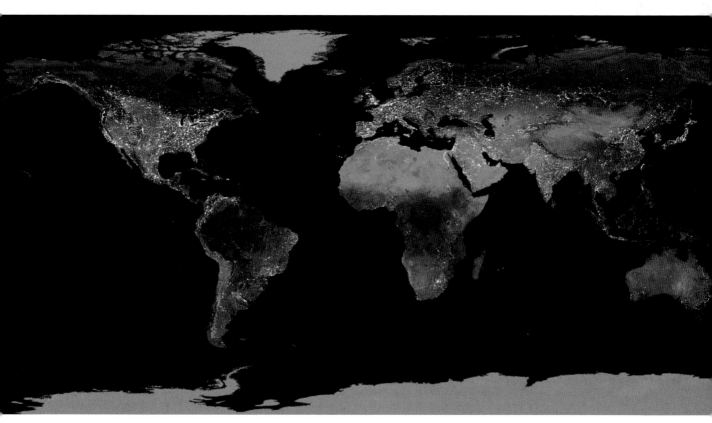

Permanent ice on land consists of glaciers in high, mountainous areas and the large ice sheets of Greenland and Antarctica. Mountain glaciers are already in rapid decline and could contribute another 20 centimeters (8 inches) to sea level before they completely disappear.

Current estimates suggest that Greenland is almost certainly adding to sea level rise, principally because its outflow glaciers are now moving faster and are delivering more ice to the ocean. An increase in surface melting is also taking place, even while snow is accumulating on central Greenland.

The Antarctic may also be losing ice mass. The dramatic melting on the peninsula and of coastal glaciers in the west is partially balanced by accumulation on the cold East Antarctic ice sheet. However, we have only a rudimentary understanding of the dynamic nature of glaciers and how the ocean interacts with the West Antarctic Ice Sheet, which sits on bedrock that would be below sea level if the ice sheet broke up. This implies that the ocean is in direct contact with the base of the ice sheet and can warm it from below at the same time it can be warmed from above. No one can rule out substantial increases in ice flow (and hence sea level) from these sources. Antarctica as a whole is not projected to warm as fast the Arctic, largely because of its greater isolation and the large thermal inertia of the southern oceans. However, warming of the surrounding ocean could affect the West Antarctic ice sheet dramatically. Therefore, although great uncertainty exists, total sea level rise could conceivably reach 1 meter (40 inches) by 2100.

A lesson from paleoclimatology is particularly relevant here. During the warm period that preceded the last ice age, around 125,000 years ago (sometimes referred to as the Eemian interglacial), temperatures in the Northern Hemisphere were not more than a degree or so warmer than today. However, sea levels were between 4 and 6 meters (up to 20 feet) higher. This water is thought to have come partially from Greenland and partially from Antarctica, though the rate at which melting occurred is unknown. This partial analog for future rises in sea level is rather dramatic. The primitive state of our ability to model how ice sheets behave means that our projections for ice sheet melt are much less quantitative than for other quantities discussed here.

However, sea level will not rise uniformly across the globe. Regional variations are determined by local movement of the land: geological uplift, such as in Sweden, which is still rebounding from the weight of the ice that pressed it down during the last ice age; or subsidence or groundwater extraction, which cause cities such as Venice or New Orleans to sink. Changes in ocean circulation will also be significant.

CHANGING OCEAN CURRENTS

Most projections of ocean circulation predict a gradual slowdown of the overturning circulation that brings a substantial amount of heat to Europe and the North Atlantic. This possibility is often discussed in rather overblown and apocalyptic tones (for instance in the film *The Day After Tomorrow*), but the consequences of the simulated 30 percent decline in the strength of that circulation by 2100, while serious, are not as dramatic as sometimes portrayed. The temperatures around the North Atlantic would probably not rise as dramatically as the rest of the world (due to the reduction in northward heat flow carried by this circulation). Local impacts on sea level rise might vary by about 10 to 20 centimeters.

In projections for the Pacific region, changes in the frequency and amplitude of El Niño events dominate the projections. Some simulations indicate a slight increase in the occurrence of El Niños, but the range of responses is very large and not much confidence should be placed in this projection.

In all projections, warming tropical oceans are an expected consequence of climate change. All other things being equal, warmer oceans will lead to more intense hurricanes; models, theory, and some observations support this basic statement (Chapter 4). However, all things will not be equal. Changes in wind patterns, El Niño frequency, or upper-atmosphere humidity will all play a role in altering the frequency and intensity of tropical storms. Current climate models cannot simulate the genesis of relatively small-scale features such as hurricanes (the size of the model "chunks" are too large), and so their projections cannot be taken at face value. The science of climate-hurricane interactions is still in its infancy.

GREENHOUSE GAS FEEDBACKS

Other amplification factors are not accounted for in the headline temperature ranges. These relate to the potential impacts of climate change on natural sources of the greenhouse gases CO_2 and methane. As the planet warms, we expect oceans and land to become less efficient at absorbing the excess CO_2 (see Chapter 6). As a result, more emissions will remain in the atmosphere, increasing the climate forcing. Increases in methane emissions are expected from melting permafrost areas and methane hydrates in the ocean (Chapter 3). Warming has the potential to destabilize these reservoirs of methane, and the amounts involved are huge. Increased emissions have already been detected locally, but judging the global impact is difficult. Models disagree on the extent to which these effects will happen, but the

These bubbles under the ice in a Siberian lake reveal the methane gas that is poised to be released as the lake ice melts. Monitoring these emissions as climate changes may give us some clues about positive greenhouse gas feedbacks in the future. © KATEY WALTER / UNIVERSITY OF ALASKA FAIRBANKS

amplifying effects of greenhouse gas changes in past climates imply that these effects are real. This is something to monitor closely.

CHANGING VEGETATION AND DECREASES IN BIODIVERSITY

Anticipated vegetation changes—trees moving poleward to replace tundra, conditions in the subtropics favoring drier savanna over forest—will also amplify temperature changes, particularly in the northern high latitudes. Tundra can be completely covered in snow during the winter, but trees and shrubs stick out, reducing the albedo of the surface and encouraging more melting.

We know that ecosystems and humans will be profoundly impacted by climate change, but the complexity underlying both ecosystems and human society makes it difficult to predict these consequences in a precise way. Chapter 5 gave a glimpse of how the biosphere already has been stressed, yet the climate-related stresses to come will be even more significant.

Projections for biodiversity are dramatic. With temperature increases of only a couple of degrees, 20 to 30 percent of the species so far assessed will likely be threatened with extinction. Some high-end estimates predict that 18 percent of the Amazon rainforest will turn to savanna by 2100. Scientists have more confidence in projected changes for the physical parameters (temperature, rainfall) than in these estimates about biodiversity, but current predictions are indicative of the scale of anticipated changes.

The "other CO_2 problem"—increasing ocean acidification—as described in Chapters 3 and 6, is not really climate related. But it will impact biodiversity in the oceans, favoring diatoms and radiolaria (which make shells from silica) at the expense of carbonate-producing organisms such as corals or coccolithophores.

RISKS TO HUMAN HEALTH

Consequences for human health are equally complex. Higher mortality from heat waves and drought could be partially offset by reduced deaths from cold snaps. Insects that transmit diseases, such as mosquitoes, which are affected by climate (particularly minimum temperatures or water availability) will likely shift their ranges and hence the potential for disease transmission. Water-related stresses in the drier subtropics will increase the potential for water contamination. Increased sea level implies more damaging storm surges, even without shifts in the number or intensity of storms, because coastal defenses can be breached more easily. Flooding is one of the most important climate-related sources of mortality, and it will increase in some places due to more intense precipitation events. However, each of these impacts will also be affected by nonclimatic factors, such as changes in the patterns of human settlement, urbanization, river and wetland management, and public health initiatives, making clear predictions in this area very difficult.

AGRICULTURAL IMPACTS

Crop production in the midlatitudes will likely increase with modest increases in temperature, making it possible to plant crops earlier. Closer to the equator, however, any temperature increases are likely to be detrimental, especially if they are paired with drought. With modest increases in temperature (less than 2°C) global net productivity is projected to increase, but with higher temperature rises it is expected to decline significantly, especially if coupled with more severe changes in rainfall.

Vineyards are often noted as being climate sensitive. The dates of the harvest and the resulting vintages depend on the weather over spring and summer, so much so that records of Burgundy harvest dates have been used successfully as a climate proxy record going back as far as the fourteenth century.

As climate zones shift, different grapes become more or less suitable to the altered conditions. Projections for Californian wines indicate that the zones for pro-

Vines from the award-winning Denbies Wine Estate in Surrey, England. This is the largest single-estate vineyard in the United Kingdom, and it produces sparkling wine in the traditional style, competing successfully against French competition in blind tastings. There are now over four hundred commercial vineyards in England, significantly more than the forty-six recorded in the eleventh-century Domesday Book, the previous high-water mark of English viniculture.
© JOSHUA WOLFE

ducing good cabernet sauvignon wines will shift north from the Napa and Sonoma valleys to Oregon and Washington. In Europe, summer temperatures will likely become too warm for the grapes needed to produce classic sparkling wines like Champagne. To many people's surprise, English sparkling wines are already winning blind-tasting competitions, and English vineyards have expanded more rapidly than they have in any previous period, including in the Middle Ages.

ADAPTING TO CLIMATE CHANGE

Can we adapt to the upcoming changes in climate? Modest changes pose the least threat, and adaptation to climate changes will be easiest in regions with easy access to resources and information. However, coping with climate change will be much more challenging in developing regions where resources are difficult to access and adaption even to today's climate is often poor. Even in rich countries, some segments of the population have a very limited ability to react to climate change.

As noted earlier, even if we act to keep atmospheric concentrations at the same level they are now, the global mean temperature will continue to increase for a few decades as a result of past and current greenhouse gas emissions and the thermal inertia of the oceans. Thus, the possible role for preventative action to reduce emissions is not to prevent all further global warming, but to minimize it. Policy makers must deal with the effects of global warming that are already in motion, while also deciding on methods to prevent further warming.

Will adaptation be enough? Almost every country in the world signed the treaty negotiated at the 1992 Rio Earth Summit, pledging to prevent "dangerous anthropogenic interference" with the climate system. But action will depend on which consequences of business as usual merit the label "dangerous". Three elements of climate change are most likely to fit that description: (1) the risk of sea level rise measured in feet rather than inches, (2) the risk of increased subtropical and continental drought, and (3) the risk that key greenhouse gas feedbacks will make it more difficult to stabilize emission concentration levels. Each of these threatens millions of people and enormous amounts of infrastructure fundamental to human society. It isn't possible to state precisely at what global average temperature these dangerous impacts will occur. But the probability of such catastrophic events increases as long as temperatures continue to rise. Adaption to continual rapid rises in sea level, for instance, is impossible without massive costs.

When projecting the impacts of global warming, it is important to be humble about what we truly understand. For instance, we understand a lot about the impact of greenhouse gases on radiation in the atmosphere. But there is much we still do not know about the details of ocean circulation, or the mechanisms by which hurricanes form, or El Niño events. Our prognosis for planet Earth comes with the caveat that, although projected consequences fit with our current understanding of climate, our understanding is incomplete. Uncertainty cuts both ways, however. The ozone depletion problem (discussed in the Introduction) provides a good example of how agreement among models does not guarantee a correct projection.

Our prognosis takes into account both predictable and unpredictable consequences. Paraphrasing a recent U.S. secretary of defense, we have attempted to describe the known knowns, and point out the known unknowns, but we cannot describe the unknown unknowns. The more the planet departs from the state that we have closely observed, the more likely it is that we will run into surprises. Those inevitable surprises are unlikely to be benign.

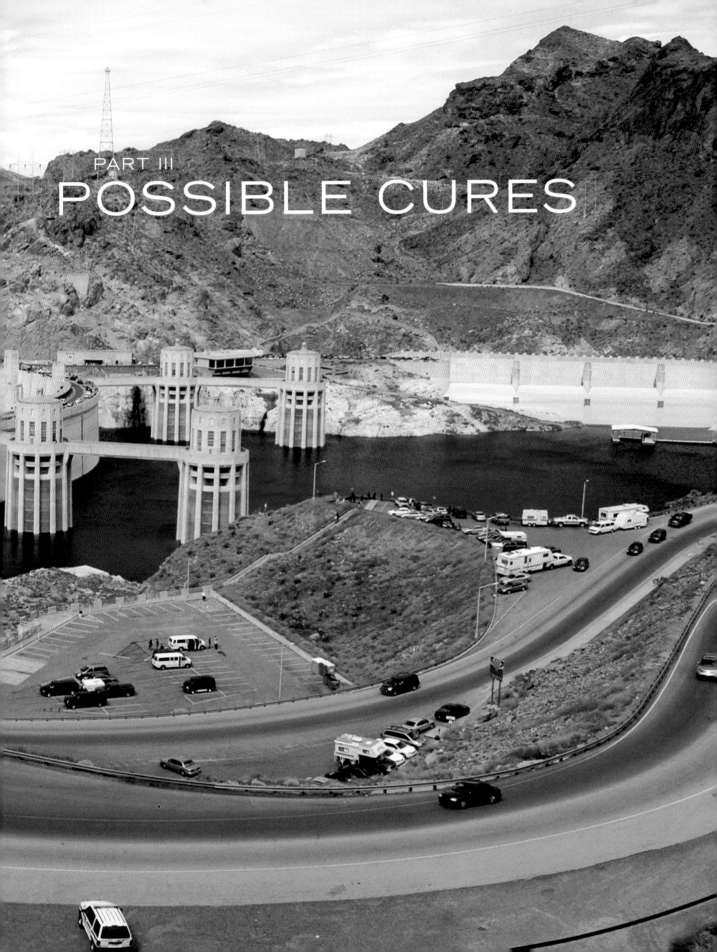

PART III
POSSIBLE CURES

CHAPTER 9

GETTING OUR TECHNOLOGICAL FIX

Frank Zeman

It is quite a three-pipe problem.
—Sherlock Holmes in "The Red-Headed League,"
by Sir Arthur Conan Doyle

There can be no doubt that Las Vegas is a monument. To *what* depends on whom you ask. Flying in, the Vegas Strip seems like a bathtub of light surrounded by the dark desert. Unseen is the faucet pouring out 250 megawatts (MW) of electricity. All in all, it seems a strange place to hold a conference on renewable energy. The 2006 Renewable Energy and Fuels Conference consisted of many people stating, in various degrees of amazement, how much growth and potential there was in the renewable energy business. Stepping outside, however, it was abundantly clear that this potential had only begun to be tapped.

Looking closer, examples of renewable energy systems are visible. The Colorado River, held back by the Hoover Dam, provides energy and water for Las Vegas and other desert cities in the southwestern United States. Hydropower is considered renewable energy based on the assumption that rain is a consistent resource. Another renewable system, based on solar heating, has been built just south of Las Vegas. Nevada Solar One is a 64-MW electrical generating station, covering 1.3 million square meters (320 acres), that concentrates sunshine for the purpose of heating oil. The hot oil then boils water and generates electricity through a conventional steam turbine. The project is feasible because Las Vegas receives an average of 6 kilowatt-hours of solar energy per square meter per day (one of the highest rates in the United States). For comparison, covering the Nevada Solar One site with 10 percent efficient solar photovoltaic (PV) panels that turn sunlight directly into

The Google campus in Mountain View, California, uses solar roof panels to generate between 5 and 6 MW of electricity per day, around 30 percent of Google's peak demand.
© JOSHUA WOLFE

213

This view of the Lake Mead side of Hoover Dam in 2005 shows the reduced water level caused by the ongoing drought in the American Southwest. The reservoir's bathtub ring indicates the normal water level. Since this photograph was taken, the lake has continued to fall, reaching about 46 percent of capacity as of late 2007. © GARY BRAASCH

electricity would generate only half as much electricity. To replace all the electricity delivered by the Hoover Dam (2,000 MW), one-third of the metropolitan area would have to be covered with the same PV panels.

In another effort to increase our use of solar energy, California signed the Million Solar Roofs Bill into law in August 2006. The legislation aims to add solar panels to 1 million houses by letting customers sell excess energy back to the grid and mandating the use of solar panels for new home construction. Regulations, or incentives, are necessary because electricity from PV panels is more expensive than fossil power. The sheer number of PV panels necessary for such an implementation is expected to drive costs down through the benefits of mass production.

Turning roofs into solar power plants is one method of reducing carbon dioxide (CO_2) emissions to the atmosphere. In general, there are three basic ways to do this: one can improve energy efficiency, use emission-free energy sources (renewable power), or prevent emissions from reaching the atmosphere (carbon capture and storage). Ways of implementing these methods are commonly called mitigation strategies. If necessary, attempts could also be made to cancel out any adverse effects of climate change through geo-engineering (the deliberate modification of the Earth's environment to affect the climate).

The term *mitigate* means to moderate or make less severe, and in this context it refers to actions that will reduce the amount of greenhouse gases emitted to the atmosphere. Any sensible mitigation strategy needs to incorporate an understanding of the entire life story of a process or product, through a technique called life cycle analysis. For climate change, life cycle analysis attempts to quantify the emissions released at all stages of a product's existence, from its production, distribution, use, and eventual disposal. This type of analysis is a useful tool for drawing comparisons between products. Let's say that you were presented with a choice between using either corn-derived ethanol or gasoline as fuel. At first glance, corn is grown, while gasoline is refined from fossil oil. However, a life cycle analysis of corn ethanol suggests that the emissions released from producing and using fertilizer to grow the corn, operating the necessary farm equipment, and manufacturing the ethanol may be similar to simply using gasoline. Recent work suggests that if you include indirect changes—things that occur as a result of using corn for fuel instead of food—it is actually worse than gasoline. Such analyses are therefore

Installation of a solar cell on the roof of a one-story home near Sacramento, in one of the first developments built under California's One Million Solar Roofs initiative.
© GARY BRAASCH

essential in determining whether an alternative is truly beneficial or merely shifting the burden elsewhere.

THE ENERGY PORTFOLIO

We use many forms of energy in a myriad of ways. Much of the energy that we use is carried through electricity and produced from a range of primary energy sources: a mixture of coal, nuclear, natural gas, petroleum, hydropower, and other renewable sources. Energy can also come directly from systems such as gas stoves, furnaces, and water heaters.

Each fuel can be defined by the amount of greenhouse gas emissions per unit of energy, and so some are already cleaner than others. The energy in coal derives from the photosynthesis of plants that lived in swamps and wetlands millions of years ago. Coal has a very high energy-to-weight ratio, but since it is almost pure carbon, the emissions of CO_2 per unit of energy are the worst of any fossil fuel. Oil and natural gas result from the decay of plankton and other organic matter at the bottom of oceans. Water pressure condenses sediments on the ocean floor, forming sedimentary rock, and it is within this rock that oil and gas can form at very high temperatures. Oil is made up of nearly twice as many water molecules as carbon atoms, so burning oil emits less CO_2 than burning coal. Natural gas is largely made up of methane, and because it contains more hydrogen (and less carbon) than oil or coal, it emits less CO_2 than either one during combustion. Natural gas, therefore, is one of the "cleanest" forms of fossil fuel. Fossil fuels supply around 86 percent of the world's energy, with coal accounting for 40 percent, oil 39 percent, and natural gas 21 percent.

Hydropower comes from the mechanical energy of moving water, and it supplies approximately 6 percent of the world's energy, with many of the major rivers already harnessed. Water is a renewable resource, and hydropower does not directly emit greenhouse gases. Indirectly, however, hydropower can also be a source of greenhouse gas emissions through the decay of flooded vegetation in the reservoirs behind its dams, and through the release of carbon stored in the soil. Large dams and reservoirs contribute about 20 percent of global methane emissions. Additional drawbacks of using hydropower include reduced silt transport downstream (which reduces nutrient levels), increased water evaporation, and interruption of fish-spawning patterns in rivers.

Nuclear fission supplies about 6 percent of the world's energy. Nuclear energy comes from a controlled chain reaction in which neutrons cause the nuclei of heavy atoms to split into lighter atoms and more neutrons, which then causes more fission. These collisions (or nuclear reactions) release massive amounts of energy. The heat

is used to drive the same steam cycle found in a coal plant. Again, no greenhouse gas emissions are directly emitted from the process. Associated concerns include waste disposal, safety, proliferation, and terrorism. Notoriously expensive due to very high initial capital requirements, the actual cost of nuclear energy cannot be known without first establishing the cost of nuclear waste disposal. Additional problems with nuclear power were exposed in France during the 2003 heat wave, when at the peak demand for electricity, a number of stations needed to be taken offline because low river levels reduced the availability of cooling water. However, new designs, such as the high-temperature, pebble bed nuclear reactors, are currently in the pilot stage (in South Africa) and are worth keeping an eye on.

Built in the 1930s, the Bonneville Dam on the Columbia River provides around 1,000 MW of hydropower for Oregon and Washington State. © GARY BRAASCH

ENERGY EFFICIENCY

Energy efficiency applies to all energy sources and will feature heavily in all mitigation strategies. It has several components. At the simplest level, it can be described

A modern compact fluorescent bulb is about five times more efficient than a traditional incandescent bulb and lasts ten times longer. It is an example of an energy-efficient product that saves money and reduces emissions. © JOSHUA WOLFE

as the portion of a unit of delivered energy that is used in the specific, desired way for which that energy is sought. Since every transformation from one type of energy to another loses some energy to waste heat, reducing unnecessary transformations is one key to improving efficiency. For instance, natural gas stoves are more efficient than electric stoves because they avoid converting fuel to heat to electricity and then back to heat again.

Our electricity is generated mainly at centralized power plants. At the plant, efficiency is defined as the ratio of electrical output to primary energy (fuel) input. Using representative values, coal plants deliver about 35 percent of the original energy to the grid, whereas natural gas plants deliver 48 percent. A nineteenth-century French engineer named Sadi Carnot discovered that the maximum efficiency of a thermal energy system depends only on its input and output temperatures. Larger differences allow the potential for higher conversion efficiencies. Current efforts in power plant engineering are therefore focused on developing materials capable of withstanding higher temperatures in order to raise efficiency.

Energy must also be transported to the end user through conduits such as pipelines and transmission lines, with some lost along the way. Lines losses refer to the amount of energy dissipated during transmission. These losses increase with distance and can range from 5 to 9 percent. They stem from collisions between the moving particles—electrons—with the atoms in the wire. In pipelines, energy must be added to overcome frictional losses, which are proportional to distance.

Energy efficiency really gets interesting once the electricity has reached its point of use. The ability to use energy efficiently is a function of both the device (such as an electric heater) and the environment (such as an insulated building). An efficient device will be designed to minimize the amount of unnecessary material and energy waste. For instance, it is much more efficient to transport fifty people in a subway car than if each of them used a car—the subway car itself is more efficient (less mass per person needs to be transported) and the subway system also reduces waste associated with idling in traffic and stop/start conditions because the tracks are not shared.

An instructive example can be taken from regular household windows. Functionally, a window is a transparent solid used in place of the outer wall. But the actual design and construction of the window makes a big difference to efficiency. In terms of construction, if a window is installed with gaps between it and the wall, then a direct path exists for heat exchange. A simple window consists of a pane of glass forming a barrier between the conditions inside and outside. Heat transfer occurs whenever a temperature difference exists across any material, in this case on both sides of the pane. This transfer happens through conduction, whereby heat is passed from one atom to an adjacent one as if in a game of hot potato. Through this process, the cool face of the window gets hotter, and eventually grows hotter than the air adjacent to it. The excess heat is transferred to an air parcel, which moves off to be replaced by cooler air. A double-paned window reduces heat transfer (improving efficiency) by introducing an additional heat-transfer step. The outer, or inner, pane functions in the same manner as above. But once the heat reaches the other side, it must be transferred to the other pane. Here we can imagine air molecules moving around and colliding with each other. The collisions result in heat transfer between molecules. They pick up heat at the hot pane and dump it at the cold pane. Efficiency can be further improved by filling the gap with a special gas, such as argon, that does not transfer heat as readily as normal air. Adding a third pane would increase the resistance to heat transfer even further.

The main advantage of energy efficiency is that we can extract more use from the same amount of primary energy (fewer resources and power plants are required to provide the same services). Improving energy efficiency is a desirable way to mitigate climate change because it can stabilize or reduce our emission of greenhouse gases without sacrificing industrial production.

Most energy efficiency measures require paying more up front for savings later on. Unfortunately, there is a psychological barrier to paying costs at once and receiving the benefits in smaller parcels over time. However, economies of scale make newer technologies more affordable as they penetrate deeper into the market. (Firms that provide the technologies find easier ways to manufacture products for a large demand pool, and they also encounter increased competition from other firms looking to profit from the market opportunity.) Thus, the attractiveness of efficiency measures tend to increase over time.

In deciding on the best energy efficiency measures, it is important to take the big picture into account. For example, increasing the thickness or gauge of copper wire can reduce electrical losses in the home, but these savings must be balanced against the energy consumed mining and refining the extra copper. Here again, life cycle analysis is a useful tool.

So how much energy can we save by improving efficiency? Estimates vary wildly. In the case of the compact fluorescent bulb, energy consumption is reduced by a factor of five (a 12-W compact fluorescent bulb produces the same illumination as a 60-W standard bulb). It may be difficult to achieve 80 percent reductions in all cases, but the example of the lightbulb shows that innovative engineering can produce strikingly effective results. For the sake of example, if we could figure out a way to reduce all of our energy consumption in the developed world by 80 percent, this improvement in energy efficiency would allow five times as many people to enjoy our lifestyle at current electricity generation levels. This number would include most of the world's current population. As impressive as that would be, we still need ways to produce more energy because global population is growing. Furthermore, power generation currently emits greenhouse gases, so efficiency alone will not be enough.

RENEWABLE ENERGY

The possibility of a global renewable energy infrastructure from solar, wind, or tidal power, has received a lot of attention in the media. These systems are considered sustainable because they produce electricity without emitting greenhouse gases or other air pollutants. However, the rate at which they produce electricity is inconsistent and sometimes difficult to predict—the so-called intermittency problem. After all, the wind doesn't always blow and the sun doesn't always shine. In the forms that they currently take, renewable wind and solar energy technologies are more difficult to manage than so-called base load plants (such as nuclear or fossil fuel power plants that can provide a continuous amount of electricity, or load). Incorporating substantially more renewable energy into a nation's energy portfolio would require designing a system that could successfully integrate variable loads (renewables) and more predictable loads (fossil fuels). At current levels, though, intermittent energy sources can easily be added to the power mix, and there is time available for engineers to develop storage and integration technologies that will be needed for future, more intermittent, generation profiles.

The common sources of renewable energy are solar energy, wind, and hydropower. As discussed earlier, hydropower is an established technology that has been deployed throughout the world. The mid-nineteenth century saw a boom in large hydropower projects that has since quieted down. In part, this is because the social and environmental impacts of large dams weren't as beneficial as projected. Hydropower is an attractive option because, while replacing fossil fuel as an energy producer, it could also provide irrigation services and flood control. Projects like the Three Gorges Dam in China will continue to increase the total energy produced by hydropower, but a doubling seems unlikely. Existing hydropower systems may prove a useful method for storing energy. During excess electricity production from more intermittent sources, water can be pumped behind the dam for later use through the turbine as if it were a giant rechargeable battery.

Given that solar PV systems are modular, there is no reason they cannot produce large amounts of energy. Indeed, a square PV array 160 kilometers (100 miles) on each side in the desert Southwest would be sufficient to meet all current U.S.

demand. However, although the growth rate of installed solar production capacity is high, around a 35 percent increase per year, the total peak capacity of all solar PV systems is still only 3,700 MW, or 0.1 percent of world electricity demand. The challenge is to install enough solar panels to significantly reduce emissions of greenhouse gases by avoiding fossil fuel plants. As mentioned earlier, we would also need to coordinate a backup power source (or find a way to store electricity) so that we could still use power during cloudy weather. At the present time, adding solar PV systems is particularly beneficial, as they produce peak power at times of peak demand—hot summer afternoons.

Wind power is another renewable energy solution to climate change. It is more commercialized and therefore cheaper than solar PV panels. However, intermittency issues remain. Additionally, wind energy systems are much larger and more visible on the landscape than solar, which has already been a roadblock in the

Modern windmills in the Netherlands generate a peak output of up to 2 MW each, but only when it's windy. The country's total capacity is around 1,500 MW (out of a total of 48,000 MW in the European Union as a whole) and is increasing at about 18 percent a year.
© GARY BRAASCH

construction of some wind farms (most notably off Nantucket Island in Massachusetts). Noise pollution, visual obstruction, and interactions with migratory patterns
of birds are also issues. Wind power is now a commercial technology, albeit supported by (intermittent) subsidies, and future developments will likely be market
driven.

Not all of the energy we consume is in the form of electricity; a significant portion is used directly for heat. Hot water heaters, cooking stoves, space-heating and
air-conditioning units could be run by thermal systems. A thermal system, such as
solar or geothermal water heaters, involves the direct transfer of heat to and from
the environment. Here we refer to simple, low-cost systems and not solar concen-
trator systems such as Nevada Solar One. Geothermal energy is mostly associated
with hot springs, which occur in seismically active areas such as the Rocky Mountains and Iceland. In tectonically active zones, geothermal heat is sufficient to produce electricity, and the leftover hot water can be used for space heating.

In Iceland, proximity to hot springs makes geothermal sources an obvious choice for Reykjavík's heating and energy needs. The Svartsengi plant converts geothermal heat to steam to run turbines. © GARY BRAASCH

On a larger scale, geothermal energy can allow us to use the Earth as a heat sink. A short depth below the Earth's surface, the temperature is a steady 13°C (55°F). If the temperature aboveground is cooler, water circulated to this depth can be used for heating; and if it is warmer above, it can be used for cooling.

Solar thermal and geothermal energy both benefit from an essentially free source of heat and a single-step conversion to usable energy. Solar thermal systems avoid the energy losses that result from electric stoves, electric water heaters, and electric space heaters, all of which require multiple conversions from heat to energy and energy to heat. They do, however, require a conventional backup (again due to intermittency).

Intermittency is the common challenge for all power systems based on natural energy flows. Human energy consumption follows a pattern that generally has a peak in the afternoon and a trough before sunrise. But we also need electricity at night and regardless of weather conditions. Natural gas turbines are a logical choice for backup given that they are an established, clean-burning, and low-cost technology. However, the cost of the backup system may deter utilities, given that they are not earning returns when the equipment sits idle. Another approach is to store excess electricity for use during low production periods in electrical, chemical, or gravitational form (such as in batteries, as hydrogen, or pumped water storage in reservoirs). A larger power grid, encompassing whole continents as suggested by Buckminster Fuller, allows the effects of intermittency to be dampened by widening the area of available renewable power systems.

For some final thoughts on renewable energy systems, let's return to the laws of thermodynamics, which crudely state that you can't get something for nothing, and in fact, you're lucky if you even get close. The production of electrical energy must result in the removal of another form of energy from somewhere else. Hydropower dams up the waterways to harness the potential energy released during elevation changes. Wind power removes kinetic energy from the lower atmosphere. Solar PV panels shade the ground below them. At the other end, wherever electricity is used, heat is released. These issues are not a current concern, but may become problematic if renewables are called upon to produce globally significant amounts of electricity. Large hydropower projects can dramatically increase local evaporation. Wind power might remove a sufficient amount of energy from the atmosphere, slowing down the wind. These changes would be a part of the environment that humanity created for itself. Given our current urban environment, solar roof panels may result in the smallest net changes. These effects of renewable energy sources may eventually need to be compared with the (usually much larger) effects of resource extraction.

CARBON CAPTURE AND STORAGE

Currently, 86 percent of the primary energy consumption is derived from fossil fuels, the source of most greenhouse gases. This fraction is rising as developing countries harness the cheapest, most abundant sources available—particularly coal. Coal is the only energy source found in virtually every country, and its processing is simple: dig, crush, and burn.

Given our current resource portfolio, it is only prudent to investigate methods for controlling the resultant CO_2 emissions. If we could capture CO_2 at its point of generation, pipe it to a storage location, and monitor it carefully so as to prevent any accidental releases, we could generate electricity with any fuel we like. The need for a pipeline suggests that this is only practical for large and immobile sources, such as industrial plants and power generation facilities. Unfortunately, CO_2 is not the only component. Before CO_2 can be captured, transported, and stored, it needs to be separated from the rest of the exhaust gases (particularly nitrogen, the dominant component of exhaust). There are three standard options for producing a highly concentrated stream of CO_2, each with an associated shorthand: remove it before combustion (precombustion), remove it afterward (postcombustion), or avoid mixing it in the first place (oxy-combustion).

Precombustion capture is a "gasification" method of converting fuel into a mixture of CO_2 and hydrogen. This process is at the heart of new technologies known as Integrated Gasification Combined Cycle (IGCC). The gases in the mixture can be readily separated, the hydrogen burned, and the CO_2 sent to storage. This technology is more capital-intensive than conventional power plants, as it is not yet widely implemented. The motivating advantages are high efficiency, reduced emission of other pollutants, and a low cost for capturing CO_2.

Postcombustion capture can be viewed as an additional gas cleanup step after the removal of other pollutants. It is well established in the refining and chemical industry and considered the benchmark technology for removing CO_2 from exhaust gases. Because this happens after combustion, it has no impact on the existing plant but does reduce energy efficiency and increases the amount of CO_2 to be stored. It is most analogous to the sulfate removal in power plants associated with the Clean Air Act.

Oxy-combustion involves burning the fuel in pure oxygen. The word *capture* doesn't show up in the moniker because it is no longer necessary. The combustion products are now mostly CO_2 and water, which can easily be separated. The difference between this and conventional combustion is the absence of large volumes of nitrogen, the main constituent of air. In standard combustion, nitrogen

The Kimberlina plant in California is a demonstration oxy-combustion power plant. Oxygen is separated from the air prior to combustion so that the fuel is burnt using pure oxygen, producing steam and CO_2, which can then be sequestered (see www.cleanenergysystems.com for more details). The heart of the technology is the oxy-combuster, which is based on techniques used in rocket engine design. © JOSHUA WOLFE

acts as thermal ballast, controlling the temperatures in the boiler/turbine, whereas in oxy-combustion recycled CO_2 from the exhaust is used instead. Oxy-combustion is potentially the most efficient process for isolating CO_2, because CO_2 no longer needs to be separated from other gases after combustion, although boilers and turbines may require some redesign for the new operating conditions.

All carbon-capture methods consume energy (back to those laws of thermodynamics!). Normally, CO_2 comprises 5 to 15 percent of typical exhaust, but concentrated CO_2 needs to be more than 90 percent pure, which requires energy. The end result is that more primary energy is required to produce the same amount of electricity. Because the use of coal results in the generation of CO_2, total greenhouse gas production will increase if carbon capture and storage is implemented. That extra production must also be captured and sent to storage.

Once the CO_2 has been captured, where do we hide it? Currently, the three feasible sequestration (storage) options are: sequestration in the Earth (geological), sequestration in the ocean, and sequestration in magnesium silicates (mineral). It is worthwhile noting that the current production of CO_2, on a per capita basis, exceeds all other commodities (though we do use more water). In other words, capturing and storing all the CO_2 produced would be one of the largest material-handling efforts ever undertaken.

The geological storage option pumps liquid CO_2 underground into porous rock formations. By storing CO_2 in this way, we hope to isolate it from the atmosphere.

Geological storage is currently thought to be the cheapest storage option and may even generate revenues by enhancing the recovery of oil and gas. Three significant demonstration projects are currently under way (in Sleipner, Norway; Weyburn, Canada; and In Salah, Algeria) and are providing operational experience. Weyburn is an enhanced oil-recovery project that has the appealing logic of putting the waste (CO_2) back where we got the product (oil). In Salah is a storage project in a producing gas reservoir. Many other oil and gas reservoirs may not be conveniently located or of sufficient capacity to store a sufficient fraction of future global emissions. The Sleipner project is using a deep saline aquifer as a place to store CO_2. These are essentially deep, underground, saltwater deposits; they are ubiquitous, hold large storage potential, and do not contain potable water. This makes them a good place for disposing of CO_2, but because they provide no other benefit other than CO_2 disposal, there is nothing to offset the costs.

We can also dispose of CO_2, either in its dissolved or liquid state, in the world's oceans. At two miles beneath the ocean surface, CO_2 would be denser than water and would form a "lake" at the bottom of the ocean. The ocean carbon cycle (described in Chapter 6) has enough capacity to eventually absorb several centuries' worth of emissions. However, using the ocean as a CO_2 dumpster comes with its own set of worries. The oceans have already absorbed an estimated 38 percent of all CO_2 emissions since the Industrial Revolution. Unfortunately, this has caused increased acidification that already hinders near-surface coral reef and plankton growth (see Chapter 3). Forcibly adding CO_2 will only exacerbate the current problem.

Furthermore, since the ocean is in direct contact with the atmosphere, the CO_2 in the ocean and the atmosphere will eventually equilibrate. How long this would take depends on where and how deep the extra carbon was dumped. Ocean circulation in the deep ocean takes about one thousand years or so to cycle. This is a rough upper limit before the ocean begins returning the extra CO_2 to the atmosphere. If we do intentionally store CO_2 in the oceans, the deeper the better.

Mineral sequestration refers to storing CO_2 gas in a solid form. This process involves the mining of suitable host rocks, usually magnesium silicates such as serpentine and olivine. The first step is to dissolve the rock using an acid. Once the acid is recovered, the solution contains dissolved magnesium and silica, which reacts with CO_2 to produce magnesium carbonate. (This process is somewhat analogous to the formation of limestone deposits on the ocean floor.) The good news is that the product converts the carbon into its most stable and climatically inert form. The bad news is that the reaction is slow and requires energy-intensive pretreatment steps such as heating and grinding. It is also the most expensive method of storing CO_2.

The process would result in a dramatic increase in mining and can only be considered feasible after all of the associated emissions have been balanced against the CO_2 sequestered. The product, huge volumes of solids, may be useful in adapting to climate change though. Seawalls anyone?

The aforementioned technologies for carbon capture and storage would increase the cost of producing electricity by 20 to 100 percent. Once you factor in fixed delivery and service costs, the average utility bill might end up rising 20 to 60 percent. Regardless of the extent to which they are implemented, these carbon-capture techniques can only address about half of the annual CO_2 emissions to the atmosphere. The remaining emissions are produced from sources that are mobile (such as cars and trucks) or too small to warrant a pipeline (such as small industrial boilers and home heating). The solutions for handling these emissions are less obvious and likely to be less economical.

SMALL AND MOBILE SOURCES OF CARBON

The search for a carbon-free fuel for the residential and transportation sectors inspires much debate. Right now it is uneconomical to produce such a fuel from renewable power, although this would be the ideal. If a battery-based transportation system is envisioned, then its "cleanliness" depends on the source of electricity. In terms of portable fuels, hydrogen, methanol, and ethanol have been touted as likely candidates for replacing gasoline. The first two would need to be produced at a central facility that employs carbon capture to prevent atmospheric emissions. Ethanol is sourced from biomass and could reduce net greenhouse gas emissions, depending on the fossil fuel emissions associated with the related farm equipment, fertilizer, and fermentation plant, as well as associated land use changes.

Detailed discussions of each fuel solution are widely available (start with "biofuel" on Wikipedia) but here we emphasize one point: none of these alternative fuels are resources. Their production will require energy, and the suitability of each in terms of climate change will depend on the net greenhouse gas emissions over their entire life cycle. For any of these potential gasoline replacements, if the total emissions after life cycle analysis are similar to the emissions associated with using regular gasoline, then benefits for climate change are minor. Another point to remember: biofuels would be in direct competition with food production for arable land, fertilizer, and irrigation water. This could lead to substantial increases in the price of food, and the recent increase in corn ethanol production may already be having such an effect. It also raises the question of who gets to decide when to produce food or fuel.

The potential gasoline replacements mentioned above, as well as more exotic ones, implicitly depend on a clean, cheap source of energy. Electricity is often envisioned as this source of energy, but as discussed earlier, its cleanliness depends on how it is produced. The transition to an emissions-free electricity sector has yet to begin, and there is no obvious favorite.

OTHER GREENHOUSE GASES

Mitigation is often focused on controlling the emissions of CO_2, but there are other anthropogenic contributions to climate change (as discussed in Chapter 6). The other greenhouse gases are present in lower concentrations, are not emitted at highly concentrated locations such as power plants, are not growing as fast as CO_2, and do not have as long a lifetime in the atmosphere—and it is for these reasons that they factor less prominently in mitigation discussions. Nonetheless, it makes sense to include other greenhouse gases in mitigation plans, because it may be possible to reduce their emissions significantly and more efficiently than CO_2.

Black carbon (soot) is another source of atmospheric forcing associated with industrial and biomass combustion (see Chapter 6). It is difficult to control emissions from forest fires or agricultural burning, but for controlled combustion (within power plants or factories) there are some

An unassuming methane-recovery well (one of 660) on the Staten Island Fresh Kills landfill site. Beneath the surface, the landfill is continuing to decompose. Methane from the interior is funneled to the wells from where it is piped to the power-generation plant on site. In addition to providing power, this system reduces the danger of landfill explosions and caps greenhouse gas emissions. © JOSHUA WOLFE

solutions. Black carbon can be catalytically converted to CO_2 by creating a solid trap in the gas cleanup process. The trap immobilizes the carbon until it reacts with oxygen to form a gas that can exit the device. Increasing the efficiency of combustion devices could also have a significant impact, particularly in newly emergent economies such as India or China, where huge amounts of air pollution (the Asian brown cloud) are already posing climate and health threats to the region.

The main sources of anthropogenic methane are the agricultural sector, landfills, coal mines, and natural gas distribution systems. Improving operating practices can reduce emissions from the latter three and sometimes generate heat and power, too. The actual mitigation technique is to simply burn the gas (that is, convert the methane to CO_2), though if it is present only in low concentrations this might be a challenge. The methane emissions from the agricultural sector stem from irrigated crops such as rice and the digestive system of ruminants (primarily cows and sheep). Modified agricultural practices, such as water aeration or drip irrigation, reduce emissions from croplands by reducing the amount of standing water in fields. Altering the digestive system of ruminants to eliminate methane-generating bacteria also has been proposed. The feasibility of this option may depend on public appetite for modifications in the food supply. If the widespread use of growth hormones is any example, then such biological fixes will likely be attempted. The timescales for all the above solutions are shorter than for CO_2, which will speed along the mitigation benefits. Methane levels have stabilized in recent years (at about twice the preindustrial level), implying that further mitigation efforts could bring concentrations down further.

Nitrous oxide is no laughing matter. It is the third most important trace greenhouse gas and 296 times more potent than CO_2 on a molecule-per-molecule basis. It is both an industrial gas (used as a food propellant) and an agricultural product from livestock and fertilizer. Agriculture is the dominant source, accounting for two-thirds of the emissions. The reduction of these emissions faces the same challenges as for methane. Here again, genetic solutions may reduce emissions from livestock, and conceivably, industrial alternatives can be found.

Chlorofluorocarbons are already regulated under legislation to reduce stratospheric ozone depletion (by the Montreal Protocols). Ozone precursors (nitrogen oxides, methane, and carbon monoxide) are controlled to limit air pollution (through clean air acts in the United States and the European Union). Their emissions are falling in most developed countries. Emissions of ozone precursors in the developing world are still increasing, but similar legislation will most likely be implemented in the near future for pollution control and public health reasons.

Passive systems can also remove trace pollutants in ways that are barely noticeable. Examples include the Marunouchi building in downtown Tokyo and the Dives in the Misericordia Church in Rome, each of which contains concrete coated with titanium dioxide. Titanium dioxide has the ability to absorb ultraviolet radiation (hence its use in sunblock), which is then used to convert ground-level pollutants into benign products. Similarly, the BASF Corporation has produced a "smog eating" coating for automotive radiators (termed Prem-Air), adopted by Volvo and Nissan, that catalytically destroys ground-level ozone. These passive systems have the ability to operate without energy inputs, but the results will be quite variable given the weather conditions and differing amounts of pollution. Their advantage lies in the fact that we don't have to actively target a source, and we can avoid tinkering

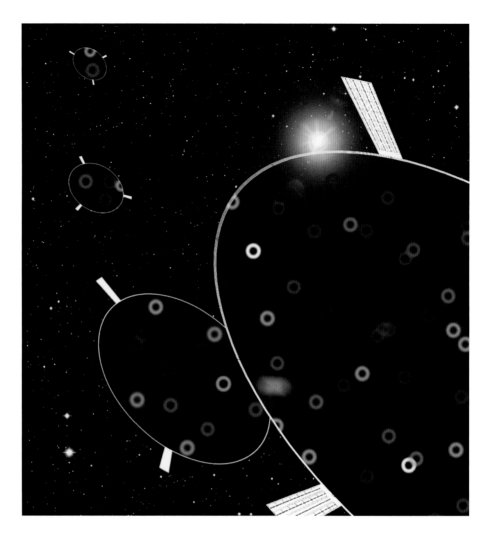

One of the more ambitious geo-engineering schemes involves placing thousands of tiny mirrors in space between the Earth and the Sun. The slight reduction in solar radiation arriving at Earth would counteract global warming from greenhouse gases. None of the scientific, logistical, financial, ethical, or legal aspects of such geo-engineering projects have yet been fully worked out.

with nature. Such a technology would be used in a "here, there, and everywhere" manner to ensure a significant pollution reduction.

GEO-ENGINEERING

The word geo-engineering is a relatively new addition to the public discourse and refers to the deliberate modification of the Earth's environment on a global scale. It has been the subject of many a James Bond film. The basic idea is to introduce a perturbation large enough to offset the potential effects of climate change. The lesson from the children's story "The King, the Mice, and the Cheese" by Nancy and Eric Gurney is useful: In that tale, a king's love of cheese leads to an infestation of rats. The rats are removed by cats, which are removed by dogs, leading eventually to lions swimming in the palace pools. In the end, things worked out for the king, but the moral is to avoid solutions that generate bigger problems. The following discussion should be read with the understanding that the moral, ethical, and political challenges of geo-engineering will likely dwarf the engineering challenge.

Humans have a long history of trying to make large-scale changes to the natural environment—for example, the introduction of kudzu in the southern United States to prevent soil erosion—with predictably unpredictable results (over 28,000 square kilometers [7 million acres] are now infested). The uncontrolled release of CO_2 to the atmosphere is, in effect, an example of inadvertent geo-engineering.

At its most basic, climate change is caused by changes in the interaction between solar radiation and the insulating properties of the atmosphere. Geo-engineering solutions would have to tackle one or both of these effects. We define as geo-engineering any process that occurs after the initial emission of the greenhouse gases.

One proposal is to introduce a structure into orbit that blocks a portion of sunlight, like a giant parasol for the Earth. Less solar radiation reaching the Earth would then counteract the increased greenhouse gas warming. Introducing aerosols into the stratosphere also has been suggested. The objective would be to mimic the effect of large tropical volcanic eruptions, which tend to cool the Earth for a couple of years afterward (see Chapter 6). One concern regarding the introduction of particles into the upper atmosphere is their interaction with the ozone layer. Extra particles provide more surfaces for ozone-destroying reactions. In both cases, the resultant climate impacts would not exactly cancel the greenhouse warming, and so regional climate change impacts would still occur.

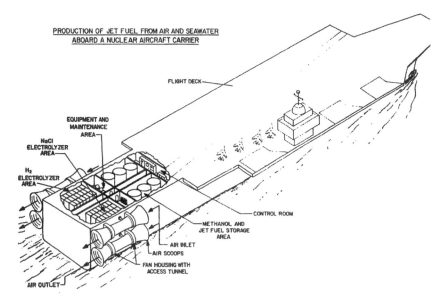

PRODUCTION OF JET FUEL FROM AIR AND SEAWATER
ABOARD A NUCLEAR AIRCRAFT CARRIER

FLIGHT DECK

EQUIPMENT AND
MAINTENANCE
AREA

NaCl
ELECTROLYZER
AREA

H₂
ELECTROLYZER
AREA

CONTROL ROOM

METHANOL AND
JET FUEL STORAGE
AREA

AIR INLET
AIR SCOOPS
FAN HOUSING WITH
ACCESS TUNNEL

AIR OUTLET

An early air capture design based on a nuclear-powered aircraft carrier, which would capture CO_2 and use excess nuclear energy to convert it to methanol for use as aircraft fuel. Although technically a renewable fuel, it relies on the existence of a secondary source of electrical power. More recent air capture designs have incorporated carbon capture and storage. © MEYER STEINBERG

A more direct way to tackle the problem is to accelerate natural processes that sequester CO_2 in the ocean or land. One idea is to increase the growth rates of plankton in the surface ocean by iron fertilization. The process is based on the fact that in parts of the ocean the limiting nutrient is iron: add more iron, extra plankton grow, and they consume more CO_2. Once the organisms die, a fraction of them fall to the bottom of the ocean, taking the carbon with them. A number of field tests have shown that fertilization does result in a plankton bloom, but the amount of carbon sequestered has been difficult to quantify. In any case, a full-scale implementation must pass the same life cycle analysis as other proposed methods. On an emissions level, the reduction caused by plankton growth must demonstrably consume enough CO_2 to significantly outweigh the emissions associated with mining, processing, transporting, and distributing the iron. Enough plankton biomass must also reach the deep ocean and not just stay near the surface. Separately, we must consider whether the enhanced "iron and plankton rain" would have any effects on other ocean life forms.

This type of tinkering with the Earth's systems or objects in space is fraught with challenges. First of all, there is no way to physically test these ideas at the global scale beforehand. Second, the global scale ensures that any implementation will be expensive. Third, it will be difficult to measure success unless done on a scale sufficient to make the effects unmistakable. Yet the lifetime of any geo-engineering technology must be short enough, say ten years, that if problems arise we are not committed to their perpetual generation. We also face the ethical and moral aspects of deciding on a technological solution that might have adverse effects on humanity, and potentially on some communities much more than others. Finally, if any geo-engineering solution is implemented, it must track our greenhouse gas emissions. If emissions continue to rise, then the scale of the geo-engineering solution must follow. This has been described as a Faustian bargain, because once started, stopping becomes very problematic.

Another idea, similar to geo-engineering in that it doesn't directly reduce emissions, is the direct removal of CO_2 from the air using photosynthesis. In its simplest form, biomass can be burned in a facility with carbon capture and storage. The traditional carbon cycle is then interrupted, and the new growth results in a net reduction of atmospheric CO_2. This can also be accomplished by storing tree trunks in abandoned mines or other similar methods. This process is limited by the rate of biomass growth, as well as the land use changes mentioned earlier.

An analogous alternative is an industrial process that actively scrubs the atmosphere of its CO_2. The technology to remove CO_2 from ambient air is not new. Its first use dates back to the early 1940s, when it was used to scrub air in a process that produced oxygen. If air is chilled to below the boiling point of oxygen (–183°C or –297°F), the oxygen will condense (like water vapor condenses on a cold window, only much colder!). However, CO_2 condenses out sooner (at –78°C or –108°F) and must be removed prior to further cooling. This idea was converted to fuel production after World War II. Fighter jets required large amounts of liquid fuel, which could be synthesized from atmospheric CO_2 and hydrogen produced by water splitting. The power needed for the cooling, splitting, and fuel production was to come from the excess power of nuclear reactors on ships.

At the turn of the twenty-first century, this concept, now known as *air capture*, has been recruited to the task of climate change mitigation. In this new manifestation, CO_2 is removed from the atmosphere, compressed, and stored using the disposal methods described above. Air capture has the unique advantage of being completely decoupled from the energy infrastructure. This means that the capture of CO_2 is not physically linked to a source, as is the case with capture at a power plant, and it provides a method for capturing emissions from hard-to-handle sources such as airplanes. To this extent, it is an alternative to the carbon-free-fuel ideas discussed above, but similarly, its actual costs and feasibility have yet to be determined. However, it will always be more expensive than the carbon capture from concentrated sources such as those discussed above. The good news is that air capture sets a ceiling on the cost for climate change mitigation. Because air capture is not attached to a specific emitter, it can be set up anywhere at any time. If any proposed emissions reduction scheme gets more expensive per ton of CO_2 than air capture, it would make economic sense to do the latter.

SOLVING THE PROBLEM
ONE WEDGE AT A TIME

Considering everything we know, it is clear that mitigating climate change is a complex problem that can seem overwhelming. There are many sources of greenhouse gases and an equally large list of potential solutions. To avoid paralysis, Robert Socolow and Stephen Pacala of Princeton University devised a method for turning "one Herculean task into several monumental ones." They considered actions that start at zero and slowly grow to midcentury. The cumulative emission reductions take the shape of a wedge. Each grouping of technological solutions, such as renewable power and efficiency, constitute a wedge. Then instead of asking how to solve the climate problem, we ask how many wedges are needed, and how much growth of technology is required for a particular wedge to be a significant part of the entire solution. This system cleverly admits to the work that lies ahead, but also shows us that solutions are achievable. Each of the elements discussed previously could conceivably constitute a wedge. The current task is to begin implementation of the most technically mature and economical wedges.

If our goal is to produce a sustainable society, then to a certain extent, we will eventually need to mimic (and possibly enhance) natural systems. This idea has been expressed through various catchphrases, such as "waste is food" or "cradle to cradle." The general idea is to move toward a zero-waste society. A flowering tree provides an apt analogy: The tree seeks to produce seedlings to replace it. In the process, it produces an abundance of flowers and seeds that interact with the whole forest. Within the forest, little of that material is wasted, and much is recycled to the benefit of the few seedlings that germinate. If we view our energy systems in this light, then the ubiquitous installment of small-scale renewable energy systems seems a promising path. Going back to Vegas, we can imagine the Luxor Hotel covered in PV panels rather than dark glass. The miniature Eiffel Tower could be coated with nickel catalyst to reduce smog and nitrous oxide. These nice thoughts could save us from more drastic solutions.

Joshua Wolfe

ADAPTATION:
FIGHTING MOTHER EARTH

Nature can be capricious. It can fail to provide water where it is expected, and deliver too much where it is not. Civilizations the world over have tried to insulate themselves from the effects of weather by using ever more complex constructions in an attempt to gain control over the climate. Sometimes they even succeed for a time. As our own actions are beginning to accentuate these climate problems, Joshua Wolfe has been documenting the scale of human response, and what we are planning for the future. However the challenge is met, fighting Mother Earth is unlikely to be quick, easy, or cheap.

Cities in arid zones around the world are experiencing freshwater shortages as a function of both changing climate and increasing demand. Desalination plants can turn seawater into freshwater, but at an enormous energy cost. For instance, the reverse-osmosis process at the desalination plant in Perth, Australia, pressurizes seawater to six times atmospheric pressure to force it through a membrane that captures the salt.

Inhabitants of coastal regions are worried about the rising tides associated with global sea level rise, which often exacerbates local subsidence. The time-honored approach is to build ever higher seawalls, dams, and levees. Nowhere is this more seriously thought about than in the Netherlands. The Delta Works project consists of a series of twelve dikes and dams that has taken forty-four years to complete, all with the aim of preventing a repeat of the disastrous 1953 North Sea storm surge that killed almost two thousand people. One of the most innovative parts of the project is the Maeslant barrier outside Rotterdam. It consists of two giant arms that extend 360 meters (almost a quarter of a mile) across the channel when needed. The Eastern Scheldt storm surge barrier near Zierikzee was the most difficult stage of the Delta Works project, involving the construction of three separate dams and an artificial island (now housing an amusement park). Environmental pressure was successful in having the plans for a full barrier changed to allow for the free flow of water to protect the local ecosystems at times of normal water levels.

On a slightly smaller scale, the Thames River barrier downstream from London was built in the early 1980s with the expectation of only having to raise it once or twice a year, but since 1990 the average rate has doubled to four times a year. Plans are already in place for an even more massive barrier to withstand further projected rises in sea level.

Subsidence in Venice (23 centimeters [9 inches] in the last hundred years) has led to frequent flood events, making the problem of rising seas there more acute. The MOSE project is currently under construction in an attempt to control flooding in the city. Sunken barriers around the Venetian lagoon will be raised to cut off the water at times of high tide. The construction of the ship canal in the Lido section of the project is expected to take over eight years to complete.

But sometimes it may be wiser to roll with the punches. Back in the Netherlands, some residents in Maasbommel are building homes that are designed to float when sea levels rise, while remaining tethered to the spot. Just in case.

The reverse-osmosis room at the Perth desalination plant in Australia

The Maeslant barrier outside Rotterdam, Netherlands

The Eastern Scheldt storm surge barrier near Zierikzee, Netherlands

The Thames River barrier downstream from London, England

Construction of a shipping canal as part of the MOSE project in Venice, Italy

An artificial island being built to bridge the 800-meter Lido inlet as part of the MOSE project in Venice, Italy

Arie and Marianne Smits standing outside their floating home in Maasbommel, Netherlands

CHAPTER 10

PREVENTATIVE PLANETARY CARE

David Leonard Downie, Lyndon Valicenti, and Gavin Schmidt

Politics is the art of the possible.
—Otto von Bismarck

Politics is not the art of the possible. It consists of choosing
between the disastrous and the unpalatable.
—John Kenneth Galbraith

What's the connection between fifteen thousand people from 190 countries flying to a resort on an Indonesian island for two weeks of meetings, Wal-Mart redesigning its truck fleet to reduce fuel costs by 50 percent, and a homeowner in Chicago buying a pack of compact fluorescent bulbs? All are partial responses to the need to limit carbon emissions. The international meetings get most of the media attention, while the more practical and smaller-scale actions being made by individuals and companies are relegated to the business pages or the last pages of books like this. But how can we tell whether these actions are sufficient or worthwhile?

The essence of the climate change dilemma, and the focus of international climate talks, is that although global community leaders are increasingly aware that climate change is a serious problem, "the blunt truth about the politics of climate change is that no country will want to sacrifice its economy in order to meet this challenge," as Tony Blair said in his speech at the 2005 G8 climate change conference. So what additional national and international policies should be enacted to address the climate challenge? How much will these cost and how does that compare with the costs of doing nothing? Who should shoulder the majority of these

Delegates at the opening session of the UN climate talks in Bali, Indonesia, in December 2007. © GARY BRAASCH

251

costs? How much and how quickly do we need to reduce emissions of carbon dioxide (CO_2) and other greenhouse gases? In this chapter we outline some of the key debates and issues, the promising mechanisms, and the role that individuals, businesses, and governments can play.

WHY IS MAKING CLIMATE POLICY SO DIFFICULT?

The task of coming up with acceptable solutions or actions to deal with human-induced climate change faces a perfect storm of factors that promote policy paralysis: a planet-wide "tragedy of the commons," very unequal national and historical responsibilities, long timescales, and the huge number and variety of greenhouse gas sources. Some of these issues are also important in other environmental problems, but in no other case do all of them come together as they do here.

It is worth exploring the concept of the tragedy of the commons. This occurs when an available common or public resource (fish in the ocean, for instance) has a limit on how much it can be used (say, the maximum sustainable catch), and a community of individuals who must decide how to use the resource (how many fish to catch). The tragedy is that the strategy most beneficial to any individual (to catch as many fish as possible) is disastrous for the community as a whole, because the limits of the resource can quickly be overcome, leaving everyone worse off. This is not just an academic thought experiment—this situation describes very well the collapse of fisheries from Newfoundland to West Africa. More generally it applies whenever the short-term benefits of an action go to an individual while the long-term costs are borne by the community. Traditional solutions to this dilemma have involved cooperation, regulation, or privatization, though these have not always been optimal, and indeed, have not always worked. They do have in common the notion that the external cost (to the community) has to somehow be incorporated into decisions about activities that cause the problem. Economists describe this as "internalizing the externalities." If this is done well, the normal mechanisms of the market should help secure the desired outcome rather than undermining it.

The tragedy of the commons clearly applies to the climate change problem—the benefit from using fossil fuels goes to the users, while the costs are paid by the whole world. This situation was described by Sir Nicholas Stern (former chief economist of the World Bank and lead author of the 2006 *Stern Review on the Economics of Climate Change)* as "a colossal market failure." That is, the situation leads to decision making based on producing benefits for small sectors while the rest of society loses out—in effect subsidizing those who need it least. The inequity is made worse because the developed nations (the European Union, the United States, Australia, Canada, Japan, and Russia to some extent) have contributed the lion's share to existing greenhouse gas burdens and benefited from the cheap energy produced, whereas the costs of climate change will likely be borne disproportionately by the poorest and least economically developed parts of the planet. There is also an inequity over time. Current generations benefit from cheap energy made possible by emitting greenhouse gases to the atmosphere, but the costs of the climate change produced by these emissions will be borne by our children and grandchildren. We benefit today while they have no choice but to deal with the world we leave them.

Another major obstacle to developing climate policy relates to long timescales. These arise from three sources: the time it takes for human society to change the way it produces and uses energy, the time it takes for the carbon cycle to adjust to the new CO_2 we are putting in to the atmosphere, and the time it takes for the planet (particularly the oceans and ice sheets) to react to changes in atmospheric composition. Each of these sources are associated with timescales of decades and longer. Power stations being built today have expected operating lifetimes of forty years, the carbon cycle can take decades to fully respond to changes in emissions, and ocean warming takes twenty to thirty years to catch up to changes in the atmosphere. These lags mean that decisions being made today will have impacts for thirty, fifty, or even one hundred years. We are building up to a situation in a few decades' time in which we will have already locked in a large amount of possibly dangerous climate change. The long time periods also make it difficult to see current changes (except in special circumstances) and to predict impacts with confidence (as discussed in Chapter 8). These timescales make it easier to ignore cause-and-effect relationships between our actions today and the impacts tomorrow. They also represent significantly longer periods than the electoral cycle or corporate business plans, making it more difficult for even the most concerned politicians and business leaders to address climate change and even easier for the less scrupulous to pass along the problem to future officeholders and executives.

The variety of activities that contribute to the problem also makes developing effective climate policy a very difficult task. Power generation, transportation, industrial production, deforestation, landfills, mining, and agriculture all contribute significant quantities of greenhouse gases to the atmosphere. Given that these sectors touch on the very heart of our modern industrial society, enacting significant change will not be easy. For contrast, consider the case of stratospheric ozone. Although protecting the ozone layer was a far more difficult challenge than often remembered, key manufacturers were able to easily develop a variety of "drop-in" substitute chemicals and processes that allowed the aerosol, refrigeration, and air-conditioning industries to reduce their emissions of chlorofluorocarbons substantially and economically in a relatively short period. Addressing climate change involves introducing potentially significant changes to a much larger, broader, and more complex network of economic activities.

A number of debates over fundamental issues also inhibit rapid action. Because industrialized countries are responsible for the vast majority of emissions of greenhouse gases in the nineteenth and twentieth centuries, should they shoulder all efforts to address the issue? Or since some large developing countries with rapidly growing economies will soon be the largest national emitters, should they also take immediate action? Are emissions per capita the most relevant metric, or should emissions be reckoned on a national or regional basis? Should emissions be immediately reduced across all sectors or only where it costs least until new technologies are available in other areas? Where should efforts be concentrated to reduce the impacts of climate change? Should we prioritize improving the ability of developing countries to adapt, or prioritize the vulnerability of ecosystems, future profits, national security and energy infrastructure development, or moral imperatives about the impact on future generations? Since the answers to these questions imply

The Amos coal-fired power plant downriver from Charleston, West Virginia, is listed by the EPA as among the most polluting and greatest emitters of CO_2 among U.S. power plants. © GARY BRAASCH

very different allocations of efforts (say, between the United States and China), they are naturally contentious, although compromises are possible.

Another obstacle is the huge group of vested interests that support the status quo to continue with business as usual. These interests extend far beyond those frequently considered in this context, such as the oil companies, to include town planners, forestry interests, trucking companies, supermarkets, airlines, and so on. In fact, pretty much everyone with existing infrastructure has short-term incentives to continue business as usual—even if they agree with the longer-term need for change. In addition, these different economic interests, and different countries, face very different "adjustment costs" for reducing emissions. Without figuring out a way to spread the costs in a way that most find fair, effective policy is likely to be blocked, regardless of how much the common benefits of such action would exceed the costs over the long term.

Finally there is the sheer magnitude of the task itself. To reduce CO_2 emissions to the levels necessary to stabilize concentrations in the atmosphere at near current levels, decreases of 80 percent or more will be necessary by the end of the twenty-first century. For the purposes of these estimates, total greenhouse gas concentrations are often given in CO_2–equivalents (CO_2e)—that is, the total amount of CO_2 that would give the same radiative effect as the mix of CO_2 plus methane plus nitrous oxide and other forcings. Current concentrations are around 440 parts per million (ppm) CO_2e from the main greenhouse gases alone, but the real net effect (including the impact of cooling aerosols) is about 380 ppm CO_2e—conveniently (and confusingly) close to the actual concentration of CO_2. This shorthand acknowledges that it is the net effect of all forcings that controls climate and provides a simplified framework for the economic discussions. As a consequence of this definition, emissions of other greenhouse gases such as methane can be expressed as an equivalent emission of CO_2. Molecule for molecule, methane is a more powerful greenhouse gas. Due to its chemistry, however, methane does not stay in the atmosphere quite as long as CO_2 (see Chapter 6). The net effect of a specific quantity of methane emitted is a combination of these two effects (the radiative forcing and lifetime) and is referred to as a global warming potential factor. The global warming potential for future methane emissions is 23, implying that 1 ton of methane emissions will have the same effect as 23 tons of future CO_2 emissions over a one-hundred-year period.

The speed at which reductions of all greenhouse gases can occur will determine the eventual atmospheric level of CO_2 and the size of the climate change to come. A number of potential stabilization targets at different CO_2e concentrations are being discussed: stabilization at 400 ppm, 470 ppm, or 560 ppm. The lower the target, the more quickly reductions would need to come about. For instance, for CO_2e concentrations to stay at 470 ppm, emissions would need to peak by about 2010 and fall by 70 percent by 2050. For 560 ppm, the emission peak is around 2020 and a 20 percent reduction needs to come by 2050. In each case, further reductions must follow. It is worth noting that, because of the long timescales of the ocean, stabilized concentrations do not mean a stable climate. Warming and sea level rise could continue for decades, if not centuries

The enormity of what is required can make small steps seem pointless and not worth the effort. These difficulties are exacerbated at the global level because many countries with very different agendas and interests must choose to cooperate. Thus, even though we know what needs to be done and have the technology to do it, it is still not clear that we will act in time.

IS THERE CONSENSUS TO ACT?

On an issue as inherently global, complex, and contentious as climate change, effective policy requires, as a necessary first step, establishing consensus among key political, economic, and scientific actors that the problem not only exists but can and must be addressed. Most scientists reached broad agreement on this issue many years ago.

However, for scientific knowledge to impact public policy in a significant way, that knowledge must become widely accepted beyond the experts. For many years, scientific arguments concerning the seriousness of the climate issue ran up against honest uncertainty regarding what was likely to happen. Such scientific debate is common and a healthy part of the interaction between science and policymaking, but it did give some observers the impression that climate change was still a speculative issue. More troubling, however, was the concerted campaign by some economic and ideological interests to use this debate to undermine, discredit, and cast doubt on what was actually well known within the scientific community—that human-induced climate change was under way and would get progressively worse unless greenhouse gas emissions were reduced. As a result, key political and economic actors in many parts of the world—and to a great extent the general public—did not believe a case had been made for urgent action to restrict greenhouse gas emissions until relatively recently.

The needed consensus has now begun to emerge. The last few years have seen a remarkable shift in attitudes by policy makers, the corporate community, and the general public. This shift reflects a decade of increasingly numerous statements issued by respected national and international scientific groups on climate change.

Previous sections in this book have highlighted much of this science. The 2007 Intergovernmental Panel on Climate Change (IPCC) report's direct statements that "warming of the climate system is unequivocal" and "there is *very high* confidence that the net effect of human activities since 1750 has been one of warming," essentially ended the debate about the reality of climate change for many in the policy and business communities. Scientists will continue to learn much more about the details of climate change, and it is hoped that this increasing knowledge will help improve policy over time. However, the consensus on the most basic issues, reached long ago in the science community, is now being matched by a consensus among opinion makers in the political, business, and religious communities.

In particular, the corporate community's awareness of climate change has risen dramatically in the past few years. Many businesses increasingly support a far more significant policy response than they did even a few years ago. This movement has been led by major corporations in the financial sector, especially the largest insurance firms and institutional investors, who see huge negative implications of ignor-

The Capitol power plant in Washington, D.C., which used to supply heat to Capitol Hill buildings, is the only coal-fired plant in the District of Columbia and is responsible for a large part of the city's air pollution and carbon emissions. Efforts to reduce the amount of coal burning have been stalled by senators from coal-producing states, underlining the difficulties in organizing federal action to curb emissions.
© JOSHUA WOLFE

ing a large rise in sea level, greater flooding, or more severe storms. However, several other factors are driving private sector action on climate change in all areas of the economy. These include: concern about the unpredictable and potentially high costs associated with future climate change; new business opportunities in clean energy, energy efficiency, and carbon trading; higher prices for oil and gas; the development of new technologies, which have reduced the costs associated with lowering carbon emissions; and efforts to stay in line with public opinion, especially in Europe. Many businesses also want the transition to a new regulatory environment to happen quickly. Because they now see climate change and climate change policy as inevitable, these businesses want the new regulatory frameworks in place so they can adjust their business and investment decisions as soon as possible.

Some significant forces in the oil, coal, and transportation industries still resist new policies, although not monolithically, and important actors in each of these sectors have switched to supporting new action. Of course, these companies, like those in all industry sectors, are pushing for policies that will benefit them, or at least cost them as little as possible. At the same time, in 2006 and 2007 alone, a large number of important existing or new business coalitions called for far stronger national and international policy measures to address global climate change. One example is the Global Roundtable on Climate Change, which brought together stakeholders from all economic sectors and all regions of the world to endorse a range of global actions. Another is the call for stronger U.S. climate policy by the major corporations participating in the U.S. Climate Action Partnership.

MITIGATION AND ADAPTATION

Even with this emerging consensus, vexing problems remain: What exactly should we do? What are our options? How much will they cost? Who should pay for it? What potential impacts can be prevented through mitigation? What will we need to adapt to?

The IPCC reports and other studies have shown convincingly that the continuing and accelerating impacts of business as usual have the potential to be far too severe to be acceptable. Continual adaptation to a changing climate—developing ways to cope with, for example, accelerating sea level rise, large shifts in rainfall patterns, and disease vectors, and increasingly severe impacts on ecosystems and agriculture—would be impossible in many cases. Even where physically possible, adaptation would come with enormous and increasing economic and social costs.

Meanwhile, climate change is ongoing and more severe impacts are likely to occur no matter what steps are taken to reduce emissions. Equally important, the impacts of climate change will fall most heavily, at least at first, on the poorest and most vulnerable people around the world. Therefore, we can not ignore strong political and moral arguments that adaptation to ongoing climate change must accompany any mitigation efforts.

Mitigation and adaptation have therefore become a standard part of most political and business thinking regarding climate change. However, there is an important asymmetry between adaptation and mitigation efforts. Adaptation, such as building higher seawalls, installing more air-conditioning, or expanding reservoir capacity, generally benefits the people who pay for it. In places with enough spare capital, investment in adaptation will likely happen without much government intervention. The benefits of mitigation efforts are much more diffuse. Lowering emissions in the United States will reduce climate impacts in China as well as domestically, but will not be effective unless adopted widely. For those reasons, the "tragedy of the commons" problem with mitigation is more acute. Hence, mitigation efforts specifically require international forums and collaboration across borders.

The lack of adaptation—even to today's conditions, let alone to the future—is a huge problem in less developed countries and in less wealthy parts of developed nations. Projects such as the Nairobi Work Programme run by the United Nations are designed to assist developing nations with improving their assessment of climate impacts and vulnerability, and to make informed decisions on practical adaptation actions. The challenge is to integrate the building of enhanced resilience to climate variability and climate change with international development and aid programs.

WHAT WILL CLIMATE CHANGE COST?

Implicit in all climate policy discussions are considerations of the costs associated with the different courses of action. How much will the impacts associated with business as usual cost? How much will it cost to reduce emissions enough to forestall the most serious threats? How much would it cost to adapt to future changes? Can these costs, which stretch over decades and almost every economic sector, be accurately calculated?

Hundreds of difficult-to-predict variables affect the economic calculations. A small sample of these include: the speed at which we can deploy current technology; the timing and cost of developing new technology; the economic and environmental benefits of deploying existing and new technology; how to calculate costs and benefits far into the future; how to assign meaningful dollar values to the variety of impacts, such as species extinction or sea level rise; future levels of global economic and population growth; and the cost of particular adaptation policies.

Biomass growth can be used to store carbon that would otherwise escape into the atmosphere. These Sitka spruce and Western hemlock forests in Oregon are among the richest in the world in soil depth and in the amount of nutrients and carbon stored in the soil and woody debris. Encouraging the growth of such forests is one way to increase the amount of carbon being stored, as long as the trees aren't later chopped down.

© GARY BRAASCH

Broader questions also arise: Are current generations responsible for past emissions? What obligation do we have to future generations? Can we put a price tag on species extinction? Despite such important practical and philosophical difficulties, however, cost-benefit studies underlie much of the policy discussion.

If we take no action on climate change, the costs will relate to adapting or failing to adapt to a changing planet. They will include the costs of failing crop yields in developing regions, freshwater shortages in both developing and developed countries, changes to subtropical rainfall patterns, health costs associated with changing disease patterns, and the direct impacts of heat and floods. Other big-ticket items will include sea level rise that threatens both major cities and the viability of low-lying agricultural areas; a potential rise in the intensity of storms, forest fires, droughts, flooding, and heat waves; and the unknown impacts of serious declines in biodiversity as a result of species extinction. Low-probability but high-impact effects, such as the risk of abrupt, large-scale shifts in the climate system, also need to be quantified.

The costs of mitigating climate change are those associated with reducing green-house gas emissions, protecting forests, and maintaining or increasing potential sinks of CO_2. Carbon sinks are the inverse of carbon sources and include reforestation and changes in agricultural practices that cause the carbon in trees and soil to be taken out of circulation. If this carbon can be permanently stored, then it is equivalent to canceling out the same amount of fossil fuel emissions.

It is often assumed that the cost of reducing emissions will result in a significant decline in economic growth. Economic growth is usually measured in terms of gross domestic product (GDP), which is defined as the market value of all final goods and services produced in a country within a given period of time. However, a high GDP is not the same as high human welfare. Many factors that enhance human well-being are not factored into GDP—quality of life, habitat loss, species extinction, cultural loss, and damages to nonmarket subsistence agriculture, for instance, do not get included. As long as we remember GDP is an imperfect and incomplete measure of well-being, however, it remains a reasonable guide, if only because we have no better, widely accepted, alternative measuring stick.

Historically, CO_2 emissions have been correlated to GDP. In much of North America and Europe, for example, carbon emissions and domestic production have increased together since 1850. With advances in energy-efficient technologies and a shift toward services relative to manufacturing, however, the link between CO_2 emissions and GDP has weakened in many countries. In the United Kingdom, for instance, GDP increased by 32 percent from 1990 to 2002 while CO_2 emissions fell 15 percent, although some of this change was due to the closure of ailing steel and coal industries. However, the stage of economic development and the standard of living of individuals in a given country play an important role in this linkage. In the developed economies, the cost of mitigation can be limited to the direct costs of implementing greenhouse gas reductions. Specific carbon-intensive sectors will bear most of those costs, whereas for the other sectors, economic benefits will occur (due to greater efficiency and in the long term a cleaner environment). In developing countries, outside help is likely to be needed to ensure continued sustainable growth and enhanced resilience.

The IPCC assessed both the costs of unfettered climate change and the costs of significantly reducing greenhouse gas emissions. The cost estimated by the IPCC for keeping concentrations in a range between 445 and 710 ppm CO_2e is between a 1 percent increase and a 5.5 percent decrease of global GDP by 2050. But this cost is relatively small compared to the potentially huge impacts of climate change if we continue business as usual. As the temperature increases, IPCC estimates of the costs of doing nothing go from 1 percent of GDP for a couple of degrees of warming to over 10 percent of GDP for a 5°C (9°F) warming. Even without consideration of serious costs of climate change, decreasing greenhouse gas emissions alone could have a net economic benefit by decreasing other types of pollution, improving human health, increasing energy efficiency, and developing new energy technologies and the jobs associated with those technologies.

Other studies come to a similar conclusion. In 2006 the British government commissioned an examination of the cost of mitigating greenhouse gases from former World Bank Chief Economist Nicholas Stern. The Stern Review concluded that the cost of stabilizing concentrations in the range of 500 to 550 ppm CO_2e to be 1 percent of GDP by 2050, whereas not acting could cause a 5 to 20 percent reduction in global GDP, with developing nations bearing even larger costs. Translating that into dollars gives the cost of these impacts as many tens of billions of dollars per year, and up to trillions of dollars annually by the end of the century. Once wider definitions of human welfare are taken into account, the incentive for significantly enhanced action is even clearer—acting to address climate change will cost far less than failing to act.

WHAT CAN BE DONE?

There may be increasing agreement that we need to reduce greenhouse gas emissions, but there is far less agreement about the specific actions we should take. Chapter 9 outlined the challenges and the technological possibilities and hurdles. Here we discuss what role policy has in moving things along.

The most obvious policy tool is the mandate. Governments could simply pass laws that force emission cuts. Such "command and control" policies could take the form of limits on emissions from all factories, power plants, and cars; requiring far higher energy-efficiency standards for transportation, buildings, machines, and lighting; mandating the use of solar panels on all public schools, malls, and other large buildings with extended life spans; forcing utilities to produce an increasing percentage of their power through renewables such as wind, solar, and geothermal; and/or prohibiting certain products or activities, such as banning construction of new fossil fuel power plants that do not sequester their carbon emissions or requiring that all lawn mowers, weed trimmers, and leaf blowers run on electricity rather than gas-powered motors.

While such straightforward approaches can be appropriate in certain cases, they have a number of important disadvantages. Chief among these is that mandates often do not account for how expensive a given requirement could be or the possibility that an equal amount of emission reductions might be available in a different area. The cost of reducing CO_2 emissions will vary enormously across industrial sectors. For example, although power stations can install sequestration equipment, the only way for airlines to reduce emissions to zero would be to shut down, because, for the moment at least, no alternative to fossil fuel–based jet fuel exists. Mandates have worked in the past—most relevantly in cutting emissions of ozone-depleting substances—but their success relies on the existence of ready alternatives that do not require large shifts in behavior.

Fortunately, other policy options exist that can lead to equally significant emission cuts, but in more economically viable ways. These include trading systems, pollution taxes, fees, subsidies, and narrowly targeted regulations. Although each could be important, we focus here on the two that receive the most attention in current public debates—emission trading, sometimes known as cap and trade, and emission taxes. Each has its advantages and drawbacks.

Cap-and-trade systems work by placing a limit on how much pollution may be generated, allocating permits that divide the right to release this pollution, and then allowing companies to buy and sell these permits, leaving the price to be determined by the market. Over time, as government reduces the number of permits, their price rises, which provides increasing incentives for companies to limit their emissions and sell their permits. This leads to technological innovations that can produce even greater reductions at lower costs. The companies that can reduce the most emissions at the lowest cost win. Thus, cap and trade is considered a market mechanism for reducing pollution, as it harnesses the forces inherent in free enterprise to the pursuit of emission reductions.

The inspiration and model for a cap-and-trade system for CO_2 was the successful use of such a system in reducing sulfate pollution and combating acid rain in the United States in the 1990s. The principal source of acid rain is the sulfur dioxide emitted from coal-burning power plants, mainly in the Midwest and Appalachia. These emissions chemically react in the atmosphere and turn to sulfuric acid (sulfate) aerosols (as discussed in Chapter 6). Rainfall in the area downwind of these emissions then becomes more acidic. Over a period of decades, acid rain started to make lakes and rivers in the Northeast too acidic for their ecosystems. Sulfate aerosols also have direct impact on human health, exemplified by the three thousand people who died in the London smog in 1952, or by the huge impacts now prevalent in China.

Actions to combat these problems started slowly in the 1970s with the first Clean Air Act and the formation of the Environmental Protection Agency (EPA). However, it was only with the 1990 Clean Air Act that the U.S. government set up a cap-and-trade system for sulfur dioxide. The government issued emission permits to the power stations based on their emissions and which added up to the overall cap. In the following years, the cap was reduced until it reached the target of half the sulfate emissions that were being emitted in 1980.

The companies covered by the regulation were required to fit monitoring equipment to record their emissions. Stiff penalties were imposed if the emissions were above the allowances that the company held. Emissions could be reduced by switching to a lower-sulfur coal, installing smokestack scrubbers, or increasing efficiency. The novel aspect of the system was that these allowances could be traded. If a company reduced its emissions below the allowances, it was able to sell those allowances to companies whose emissions were going to exceed theirs. Reducing emissions was therefore worth money, while excessive pollution cost money.

How much this saves or costs depends on the price of the allowances, which is set by the market. A producer whose emissions are going to exceed their allowance must either buy more allowances or invest in technologies that reduce emissions. Only if the price of allowances is less than the cost of making the reductions are they worth buying. Therefore, the price should settle at around the level that the most efficient reductions can be made. As the cap is lowered, demand for allowances will rise and so will their price, increasing the incentive to cut emissions for those who can.

In the case of sulfur emissions, this scheme worked far better than expected. Sulfur dioxide levels fell to 40 percent of 1980 levels by 2002 (eight years ahead of schedule), with costs to the polluters that were a quarter of what was initially forecast by the EPA (and many times less than the highest industry cost estimates). The various health and environmental benefits of this cut are estimated to have exceeded the costs fortyfold.

The main difficulty in establishing a cap-and-trade system lies in allocating the permits so that the price remains relatively stable. A predictable price over time allows emitters to plan ahead and decide whether investing in equipment that reduces emissions would be worthwhile or whether they should purchase emission credits. If the prices are too volatile, such long-term investments just don't happen.

An example of this occurred in 2005 when the European Union launched its Emission Trading Scheme (EU ETS), the world's largest multination carbon-trading program. This program regulates the market for emissions from large facilities such as industrial factories and power plants and includes almost half of Europe's CO_2 emissions with, at last count, participation of twenty-five of the EU's twenty-seven member nations. Initially, the EU overestimated how many carbon allowances would be needed and issued too many. Fewer companies and utilities needed to buy them than anticipated, leading to a collapse in permit prices. But this process has since settled down and established a firm carbon market in Europe, with carbon allowances now trading at about $20 per ton of CO_2.

An important decision is whether to issue permits based on past emissions (as in the European carbon market) or to hold an auction. Issuing permits is equivalent to subsidizing the polluters because they are essentially being given free money. Auctioning permits raises money for the government that can be used to offset higher energy prices for the hardest hit, but this method burdens industry with very large up-front costs. A mix of approaches is also possible.

It is worth noting that trading schemes that appear economically efficient on paper can fail to deliver the anticipated emission reductions. Such failure can result from faulty pricing or emission allocations, inadequate monitoring, poor enforcement, or the creation of perverse incentives that end up rewarding counterproductive behavior.

One example of how things can go wrong involves HFC-23 (also known as trifluoromethane, Freon 23, CHF3, or Fluoroform). One of the most abundant of the hydrofluorocarbon (HFC) gases, HFC-23 is a by-product of the manufacture of Teflon and is also widely used both as a refrigerant and in the manufacture of semiconductors. HFC-23 is also a powerful greenhouse gas with a global warming potential of 11,700, and thus is regulated under the Kyoto Protocol. The cost of reducing HFC-23 emissions is relatively small and significantly less than reducing an equivalent amount of CO_2. Manufacturers whose factories lower their HFC-23 emissions are able to sell their unused emission reduction credits at a significantly higher price than the reductions actually cost. Sale of unused emission credits has now become a large part of the revenue for these firms. In fact, the profits on the HFC-23 permit sales make it worthwhile to build even more factories that generate HFC-23 emissions just so that they can be lowered again! The net effect is just a transfer of funds from CO_2 producers who need to buy the allowances to the HFC-23-producing factory owners. Little or no reduction in net greenhouse gas emissions results from such unintended consequences.

Emission trading schemes can be very effective provided they are designed and operated effectively. Effectiveness can increase when the target emissions are easily measured and quantified, when they can be traded efficiently between companies and even across sectors, and when they can be priced appropriately. It is also important to create monitoring and enforcement procedures to prevent cheating and provide confidence that the market is fair.

The chief market-based alternatives to emission cap-and-trade schemes are pollution taxes—that is, a tax or fee placed directly on the undesired emission. Greenhouse gas emission taxes are thus direct payments by polluters to the government, based on the amount of gas released. Carbon taxes are similar levies but are paid according to the carbon content of the fuel being consumed. Where carbon taxes have been successfully applied (for instance in British Columbia), the "carbon cost" of production is passed on by manufacturers to consumers in the form of price increases. In theory, the higher prices encourage companies and consumers across the economy to use energy-saving technologies and begin shifting from carbon-intensive fuels such as coal to less carbon-intensive fuels such as natural gas to alternative energy sources and higher energy efficiency. The companies that can provide clean energy to consumers at the lowest cost win more business. The consumer that saves more energy, saves more money.

These type of green taxes have several advantages. Most obvious are the strong incentives to reduce emissions and use cleaner energy. In addition, revenue generated by the tax can be used to spur investment in new or existing technologies and to reduce other types of taxes, such as income or sales taxes, leading to little or no practical increase in the overall tax burden on the economy.

Given the encouragement that carbon taxes could give businesses to increase their energy efficiency and the potential ability to offset other taxes, some analysts argue that such policies might actually benefit the overall economy. Furthermore, because the externality (in this case carbon emissions) is taxed directly by a known rate, the incentive to reduce emissions is permanent, making it easy to plan for the long term. The straightforwardness of the approach also discourages behavior by firms that could exploit loopholes. Lastly, green taxes are relatively easy to implement.

On the other hand, green taxes have certain drawbacks. Most important, it is impossible to know ahead of time the exact reductions that the tax will bring. While emissions trading sets the limit on pollution levels and allows the market to set the price, taxes set the price for emitting CO_2 but leaves it to the market to determine how much is emitted. It is possible that firms and individuals will simply pay the tax and continue the emissions. Conversely, too much spending could occur on goods that receive special tax preferences, leading to overinvestment in efficiency or renewables that the market is not yet ready for. Also, by driving up energy costs, CO_2 taxes could initially harm overall economic activity that would be difficult to counteract via spending the tax revenue or offsetting existing income or sales taxes.

Some ideas have been proposed to address these concerns. Taxes could be introduced slowly to minimize disruption, and the revenue could be used to reduce sales or income taxes on low-wage earners (which are very regressive). The most effective set of policies would likely combine emission trading for appropriate industries, carbon or emission taxes for other sectors, mandated regulations on certain activities, and subsidies to help develop and deploy promising new technologies.

One further set of problems arises when dealing with carbon sinks. If emitting carbon has a price, surely taking carbon out of the air should be worth the same thing? Unfortunately, many of the schemes to enhance carbon sinks are very difficult to price fairly. Exactly how much carbon has been stored in a replanted forest? And how long will it stay stored? There is also the potential to create more perverse incentives, such as encouraging people to chop down forests in order to claim a credit when replanting it. Therefore, for the time being at least, most carbon sinks are not included in tax or trading schemes.

The debate among advocates and detractors of these two approaches has focused on the environmental effectiveness, international compatibility, economic cost, administrative transparency, social equity, investment certainty, and political acceptability of each option. Many twists on these ideas have been proposed to blunt previous criticisms—"safety valves" on the market for permits that prevent the price from going too high, for instance. However, examples of each scheme are now proceeding in many regions. For instance, British Columbia has instituted a carbon tax starting at $10 per ton of CO_2 (which corresponds to about 9 cents per gallon of gasoline) that will increase to $30 a ton in a few years. Norway has instituted a very effective carbon tax as well. The Regional Greenhouse Gas Initiative in six Northeastern states is a U.S. emission trading scheme that will begin operation soon, similar to the Western Climate Initiative in Canada. With these experiences, the schemes and taxes will likely become more effective in the future.

There are many other places where government action could be effective. In particular, energy efficiency presents a win-win scenario for the economy and the environment, helping to moderate both energy demand and greenhouse gas emissions. Because increasing efficiency saves money in any case, there is already a market incentive to be more efficient, but the fact that inefficiencies still exist implies that more can be done to promote it. Often there is a lack of sufficient initial capital for investments that would pay for themselves in a number of years. Government policies can provide incentives to invest in efficient technologies or to penalize the use of inefficient products. Establishing technology standards that require certain products to have a minimum level of energy efficiency and mandating the use of environmentally sustainable technologies can also be useful.

In the area of power generation, mandating the fraction of renewable energy and allowing customers choice in their suppliers can provide further incentives for investment in clean technology. Regulatory environments can also be changed. Currently, state regulators are often forced to accept the lowest cost option for

The "green roof" atop Chicago's City Hall is aimed at minimizing urban heating to reduce air-conditioning bills in the summer and heating bills in the winter. The roof features over 20,000 plants and more than one hundred plant species. In summer, it can be up to 44°C (80°F) cooler than the roofs on surrounding buildings. © JOSHUA WOLFE

new generating capacity, but those costs may only be based on current conditions. Because it is likely at some point in the near future that carbon emissions will have a cost, a carbon-emitting plant that looks cheap now may end up being much more expensive over its forty year lifetime. It is up to regulators to make sure that forward-thinking proposals do not get rejected out of hand.

Urban planning—particularly improvements to mass transit, traffic management, and zoning for higher population density—can lead to more efficient commuting and reduced emissions. For instance, the congestion charge for cars entering central London has led to a big reduction in smog and traffic and increases in bicycling and mass transit use.

Buildings in big cities such as New York are a huge source of emissions, and buildings overall are 30 percent of national emissions. Therefore, building codes for insulation, ventilation, and water management can play a big role in making buildings more efficient. They can encourage intelligent elevators in new skyscrapers that make trips faster and save energy, or promote green roofs to reduce heating costs in winter and excessive urban heating in summer.

INDIVIDUAL ACTION

In the face of these problems, what can individuals do? Although long-term effective action will definitely require new global policies and market incentives, individuals can play a significant role in initiating and building greater support for the required changes.

The most important task is to educate yourself and those around you. Climate change may be a very complex issue, with a host of important scientific detail, but it is also one that is surprisingly understandable in general terms. The more each of us knows, the more likely we are to make correct decisions. The resources listed in the Further Reading section are a great place to start.

People are often surprised at how easy it can be to reduce their carbon footprint, which is a measure of the CO_2 emitted through the use of fossil fuels and other activities in our daily lives. It is determined by calculating not only an individual's direct emissions from air and car travel, heating, and electricity but also covers the indirect fossil fuels required to produce and transport food, clothing, and other consumer products. The carbon footprint for the average American is about 20 tons CO_2 per year, whereas the global average per person is a little less than 4 tons CO_2 per year. Depending on where and how you live, however, your footprint can be significantly larger or smaller. The carbon footprint of New York City residents is around 7 tons CO_2 per year, mostly because of the popular mass transit system and relatively high-density living.

Increasing energy efficiency in your home is the best way to reduce your carbon footprint while at the same time reducing your heating and electric bills. Although some options cost money initially, they more than pay for themselves over time through lower energy bills. For instance, improving home insulation can save the average homeowner up to 25 percent on their heating bill and reduce emissions by almost 1 ton CO_2 per year. Choosing the best energy-saving options when you

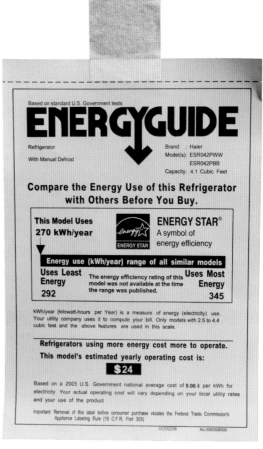

An EnergyGuide label is required on all major appliances sold in the United States to give consumers better information on energy usage. The EnergyStar logo on this label designates the model as one of the most energy efficient: it uses less than half as much energy as a refrigerator manufactured before 1993.
© JOSHUA WOLFE

replace old windows can have an even larger impact. Installing a programmable thermostat, or adjusting it down 2 degrees during the winter and up 2 degrees in summer can save another ton of CO_2 and about $100 per year.

In 2006, Americans purchasing energy-efficient Energy Star appliances—everything from refrigerators to air conditioners to exit signs—saved enough energy to avoid greenhouse gas emissions equivalent to those from 25 million cars, all while saving $14 billion on their utility bills. Another easy conservation measure is to turn off televisions, stereos, and computers when they are not in use. When these items remain in standby mode, they can use 10 to 60 percent of the power they do when on. Sensors and timers that turn off lights and other electrical items when not needed also save money. One of the easiest and most important energy-efficiency steps is to start using compact fluorescent lightbulbs (CFLs). These bulbs use about 80 percent less energy than standard incandescent bulbs and last up to ten times longer, saving about $30 or more in electricity costs over each bulb's lifetime. If every American home replaced just one old-fashioned lightbulb with a CFL, it would save more than $600 million in annual energy costs and reduce emissions by 3 million tons CO_2 per year.

As awareness of where energy comes from has increased, many electricity suppliers have started to allow their customers to choose their electric supply solely from renewable resources—usually wind, solar, or hydropower. This option can cost a little more than the regular mix (though costs are falling), but it sends a clear signal to the power companies that the demand for cleaner energy exists. Costs also are dropping rapidly on domestic solar power systems that provide hot water and offset energy needs. In some areas, homeowners can actually produce more energy (from solar and wind) than they need during certain times of the day, and can sell the power back to the grid, effectively making their electric meter run backward.

Transportation options are also a good place to make emission-reducing choices. For instance, replace as much car travel as possible with mass transportation. Biking and walking are far healthier than driving, both for you and the environment. For people who do not have a choice about driving, using a more fuel-efficient car is helpful (and that doesn't necessarily mean a hybrid). Until such a purchase is possible or necessary, however, even taking regular steps to keep your tires adequately inflated can save 0.125 ton CO_2 per year, and checking your car's air filter monthly and changing it as necessary can eliminate another half ton.

However, the most significant role for individuals is to let your voice be heard. Politicians at each level of government—local, state, national, and international—need to know that this issue is important to you and that you want them to focus on

In McCamey, Texas, the wide open spaces previously populated only by nodding donkey oil pumps turn out to be suitable for wind turbines, too. The state has the largest capacity and the fastest growth rate in wind-based energy generation in the United States. © JOSHUA WOLFE

the problem. The same is true for the media and businesses. Write letters in support of actions to reduce emissions; ask for information on how climate change will impact your communities and what action is being taken; support green-building codes; vote with your pocketbook by buying energy-efficient products, locally produced goods, and products made from recycled materials; vote for politicians that have demonstrated leadership on climate change; support local, national, and international efforts to protect habitat and plant trees; and urge towns, schools, and businesses to embrace energy efficiency and increase their use of solar, wind, and geothermal power.

INTERNATIONAL COOPERATION

Climate change is an inherently global issue. Carbon dioxide emitted in Beijing has the same climate effect as CO_2 emitted in Ohio. Moreover, all developed countries and many developing countries must all act to restrict emissions (and deforestation) for the problem to be addressed. No single country or region can address the problem by themselves. Indeed, if any try, they could be at a disadvantage, at least in the short run, in global economic competition due to increased energy costs and other measures. Only if all major countries adopt similar measures would this valid concern be alleviated.

The framework for global policy is the United Nations Framework Convention on Climate Change (UNFCCC), which was finalized as part of the 1992 Rio Earth Summit. Ratified by over 180 countries, including the United States, the UNFCCC commits governments to stabilizing "greenhouse gas concentrations in the atmosphere at a level that would prevent dangerous anthropogenic interference with the climate system." Yet this laudable objective is also nearly meaningless. Everyone can agree to it, as it lacks specifics regarding what should be done and when. Negotiators could not agree on exactly what those levels are and what should be done to avoid them, so that was put off for future discussions.

Subsequently, a long series of difficult negotiations produced the 1997 Kyoto Protocol. This treaty, and add-on to the UNFCCC, mandates that developed countries reduce their emissions to 1990 levels by 2012. Although enough countries ratified Kyoto for it to become legally binding, the United States (which contributes over 20 percent of global greenhouse gas emissions) has so far refused to join—the only major industrialized country to remain outside the treaty (Australia ratified in 2007).

Negotiations for further reductions in the post-Kyoto period (2012 to 2020 or 2030) have been under way since 2006, including the Bali meeting in December

2007, and will likely continue until 2009. The aim of these talks is to bring all the major emitters into the framework in some fashion (including India, China, and the United States). Issues under negotiation include what overall targets the world should aim for; how fast reductions will occur; how the treaty will count different types of reduction and related efforts (for instance, reducing greenhouse gas emissions, preventing deforestation, and planting new forests); what types of countries (developed and developing) will be required to take what type of actions, and who will pay for them, particularly in poorer parts of the world, where many believe emission controls should not impede development efforts. Also being put into place are mechanisms and institutions to help reduce the costs and to regulate these efforts, including a global trading market for permits, and methods to certify that projects do indeed make the reductions they claim.

Even while these international agreements are being negotiated, national, regional, and industrial efforts are also under way. Among the largest emitters, the EU has taken a clear leadership role in climate policy. The EU is close to adopting, at least unofficially, the goal of keeping global warming below 2°C (3.5°F) above preindustrial levels (remember we are already 0.8°C above as of 2008). The central policy proposal, which has not yet been formalized into law, is for EU countries to reduce their collective emissions to 30 percent below 1990 levels by 2020 if other developed countries adopt reductions, and 20 percent if others do not, with 20 percent of their energy needs being

Delegates at the final session of the 2007 Bali climate change meeting wait for leadership. © GARY BRAASCH

met from renewable sources. Further proposals are to achieve a 60 to 80 percent reduction below 1990 levels by 2050. Europe is already committed to an 8 percent cut from 1990 levels by 2012 as part of its requirement under Kyoto. As a whole, the EU is quite close to matching its Kyoto targets—mainly because of significant reductions in the United Kingdom and Germany, which compensate for increased emissions by the more rapidly growing (but smaller) Spanish and Irish economies.

Action in the United States is being led at the regional and local level. For example, six Northeast states have set up their own carbon-trading mechanism; dozens of other states have set or are considering serious policies to reduce greenhouse emissions; and hundreds of city mayors from Burlington to Seattle have pledged to meet the Kyoto targets in their own neighborhoods. Dow Chemicals, Dupont, and Wal-Mart all have announced significant cuts or plans to cut emissions, mainly through increased energy efficiency, and other companies, including giants such as General Electric as well as small companies such as solar firms, are developing new products and technologies to meet essential consumer and corporate needs while using less energy and emitting fewer greenhouse gases. Wal-Mart, for example, aims to reduce energy use in their trucking fleet by up to 50 percent, increase the energy efficiency of new stores, reduce the packaging in the products they buy from suppliers (which reduces the energy used in manufacturing and shipping them), and promote a huge increase in consumer use of CFLs. As part of their transportation initiative, Wal-Mart plans to fit auxiliary power units to their trucks to power low-level operations at night (such as air-conditioning) as an alternative to using the main engine, which burns more fuel less efficiently to achieve the same result.

The U.S. federal government has taken relatively little action to limit CO_2 emissions nationally other than to support a variety of voluntary actions. The Bush administration was active in reducing other greenhouse gases, in particular supporting the Methane to Markets partnership that facilitates methane capture from mines and landfills for use as a fuel source. The administration also took an international leadership role in efforts to accelerate the phaseout of hydrochlorofluorocarbons (HCFCs) under the Montreal Protocol, which will have a significant climate impact as well. The 2007 Energy Bill increased the automobile mileage requirements (CAFE standards) and mandated significant long-term efficiency increases for appliances, lighting, and other efforts. However, the Bush administration also challenged lawsuits to get the EPA to start addressing CO_2 under the Clean Air Act, blocked attempts by California to set new emissions standards, refused to seriously engage Congress on any national emission-reduction plans, and did not join Australia, Europe, Japan, New Zealand, and Russia in accepting binding limits on greenhouse gas emissions under the global climate regime.

At the Bali climate negotiations in December 2007, the delegation from Papua New Guinea expressed the frustration felt in much of the international community at the lack of U.S. support for stronger global policy. Addressing the U.S. delegation after it had refused to join other delegations and agree on a revised text charting a path forward toward a new climate treaty in 2009, the representative from Papua New Guinea stood in the packed plenary hall and said, "We ask for your leadership, but if for some reason you're not willing to lead, leave it to the rest of us. Please get out of the way." The room erupted into applause. Soon thereafter the U.S. delegation agreed to the final text, and the Bali Road Map was gaveled into agreement.

It is too early to know if this is a sign of more important compromises to come. The United States is hardly alone in opposing strong binding controls. China and India have been steadfast in opposing economy-wide limits on their emissions, arguing that they will act only after an extended period of action by all industrialized nations, including the United States. Meanwhile Saudi Arabia leads a coalition of oil-exporting nations that from all appearances work diligently to delay any additional global policy efforts that would reduce the ever-burgeoning demand for oil. However, the global prominence of the United States, its giant economy, and its historical and current level of CO_2 emissions, gives it a unique visibility and, some would argue, an important responsibility.

A brighter future may be ahead for climate policy. The Bush administration belatedly increased its engagement in national and international climate discussions; many climate-related bills are moving through Congress, most of which mandate 50 to 80 percent reductions in CO_2 emissions by 2050; religious leaders and congregations are engaging the issue; business and industry views are changing; and the prospects are good that the new administration will support U.S. action to mitigate climate change. Whether the appropriate policies will be chosen, and whether they will be chosen in time, remains in question.

Johannes Loschnigg

THE CLIMATE ON THE HILL

As a climate scientist working on Capitol Hill, one of the most frequent questions I encountered—not just from other congressional staff, but from anyone in Washington, D.C.—was something to the effect of "Is global warming for real?" When I tell this to most of my colleagues in the science community, the response I get is usually some level of surprise. But once you have been in Congress for a while, you begin to understand why this has been the case.

Capitol Hill can be a very reactive place. Members of Congress have so many issues they have to keep up with in their already overscheduled days that often—unless there is a vote on a particular issue or until something comes to a crisis point—congressional members may only be marginally aware of the details surrounding a particular subject such as climate change (Representative Rush Holt, one of the few scientists in Congress, likened working as a congressman to being a television set when someone is changing the channels every few minutes).

Until early 2007 there had been relatively little substantive discussion about climate change in Congress. Aside from a few interested members holding the occasional committee hearing on the subject (much of the time such hearings were attempts to show that global warming is a hoax), the only discussion that had occurred was in 1997 when the Senate unanimously passed a nonbinding resolution stating that the United States would not enter into any international treaties on greenhouse gas reduction without the participation of "developing nations"—basically India and China. For years this resolution gave many the general impression that Congress had no interest in dealing with the issue.

One of the things that begin to change the visibility of this issue—at least in the Senate—was the 2003 vote on the McCain-Lieberman Climate Stewardship Act, a bill to reduce U.S. greenhouse gas emissions. For many in the Senate, this was the first time they had paid close attention to the subject, so they had to get up to speed quickly—hence the increasing frequency of the "Is it real?" conversations. But with the leadership of the Senate (and House of Representatives) at the time, many could afford to vote in support of the bill to show their environmental concern, while knowing that in reality its chances of actually becoming law were slim.

One thing I began to notice during these years is that what people mean by "global warming" could be very different, depending on what they had been hearing or reading. Most scientists might assume that the general understanding of the layperson would closely resemble the conclusions of the UN reports on climate science: that the rise in global average temperatures of the last few decades is "highly likely" to be a result of human influence. But because of the widely varying media coverage on the subject, and the influence of certain groups and individuals on Capitol Hill, the understanding of what climate change means was very muddled. Exaggerations or misleading statements related to climate science occurred frequently and in both directions—on the one hand, overstating the impacts of climate change; on the other, completely denying the commonly accepted science.

For example, after the very active Atlantic hurricane seasons of 2004 and 2005, increasing attention was paid to the question of whether hurricanes are becoming more intense as sea-surface temperatures are rising. Because of the high visibility of hurricane impacts, what is currently a vigorous and ongoing debate in the scientific community spilled out into the public arena, with many scientists wanting to claim that their particular research (either showing a strong link between global warming and hurricanes, or no link at all) was the final word on the subject.

Conversely, for a long time many members of Congress were focused on one particular study related to paleoclimate, which purported to show that Northern Hemisphere warming in the late twentieth century was highly anomalous compared with the previous one thousand years. Many felt the prominent visibility of this study made it a symbol of the issue of climate change. Attention then focused on the flaws in the study's methodology, based on the assumption that this study was the most important issue in climate science and that pointing out any shortcomings would show the lack of scientific certainty about the larger issue. Because of this distraction, my time on the Hill was often spent deciphering what could be very esoteric scientific concepts to staff or even members (you should see what people's faces look like when you say the words "principal component analysis"!).

With the change of leadership in both the Senate and House of Representatives in early 2007, this situation is now almost completely reversed. It's as if the floodgates have been opened: the issue of climate change is being discussed almost daily in hearings; numerous pieces of legislation are tackling the question of what to do on a national level to reduce greenhouse gas emissions; and votes on some bills are likely to occur in the near future. But the issue of what to do is actually far from settled. As my former boss David Goldston, chief of staff of the Science Committee under

Sherwood Boehlert, described the situation, Congress usually has to work for years before such major pieces of legislation become law. And because of the lack of debate over (and therefore knowledge of) the climate change issue during the past decade, the reasonable discussion of what to do—How much to focus on mitigation versus adaptation? What is the best way economically to reduce greenhouse gas emissions?— could take quite a while. As Mr. Goldston puts it, "The complexity of that policy discussion will make the previous congressional debate over whether climate change even exists seem like child's play."

A FINAL NOTE

Gavin Schmidt and Joshua Wolfe

Human beings, who are almost unique in having the ability to learn from the experience of others, are also remarkable for their apparent disinclination to do so.

—Douglas Adams

The fact that many of the timescales involved in climate change are measured in decades is both a blessing and a curse. A blessing, because it means that responses to the problem can be flexible and measured, but also a curse, because we are storing up problems that it may be too late to prevent by the time they can no longer be ignored. This time lag leads to an intergenerational imbalance between the benefits of burning fossil fuels and their costs to the environment, which, in our opinion, is a real challenge. Is it ethical to bequeath to our children and grandchildren environmental problems that they can do nothing about? The argument often is made that future economic growth will enable them to pay for the fixes, but this supposes that a finite amount of money will be enough and that the problems don't continue to get worse (even assuming optimistically that growth is not adversely affected by the consequences of climate change). This approach entails risks to the future of human society that we do not consider prudent.

In this book we have eschewed the polemics often associated with this issue in favor of a "warts and all" exposition of what we know, what we don't know, and what is already being seen. Dealing with this information in a responsible way is undoubtedly difficult. Given that a ton of carbon emitted from a power station in China has the same effect as one emitted from cars in the United States or one emitted from a burning forest in Indonesia, no single all-encompassing solution can exist. Instead, multiple incremental efforts at many different levels need to be tried, some of which will be more successful than others. To maximize the innovative energies that need to be directed to solving these problems, it's clear that a price must be attached to carbon (and other greenhouse gas) emissions. This scheme also must actually succeed in reducing emissions rather than simply shifting them elsewhere. It should not disadvantage the most proactive countries, nor curtail the development of poorer countries. For all these reasons, it seems essential that an international agreement form the framework for these changes.

To end on a positive note, the political climate has changed dramatically since we first embarked on this project. In October 2005, we mounted the photographic exhibition that eventually led to this book in a tiny gallery in Brooklyn, New York. At that time, the discussion was very different. It was before *An Inconvenient Truth* had been released and before the latest round of special "green" reports appeared on major networks and in national magazines. We focused the exhibit on showing people that climate change was really happening. Now, just a few years later, it has become easier to delve into the details of the science and, crucially, what to do about it. Public appreciation of the issue is certainly greater than it was, and many have shown a hunger for more information that is both surprising and gratifying. Politically, the outcome from rounds of international climate negotiations, although not decisive, seem to be heading in the right direction, albeit very slowly.

We structured this book around a medical metaphor of symptoms, diagnosis, and possible cures. Our prescription for the problem can rightly be debated, but as Nobel Laureate Sherwood Rowland said (referring then to ozone depletion), "What's the use of having developed a science well enough to make predictions if, in the end, all we're willing to do is stand around and wait for them to come true?"

ACKNOWLEDGMENTS

Literally hundreds of people have contributed in various ways to the creation of this book—from outlining the initial concept, to helping us understand some of the details that are far from our specialties, to fact-checking the text (though any remaining errors are our own). In addition, there were the people who let the photographers crash on their floor, tag along on trips, and acquire the images for the book. Although there is not enough space to list everyone, we offer our heartfelt thanks to everyone who donated their time, energy, and resources to this project. We'd like to thank all of the chapter authors, who put up with suggestions and edits, and all the contributors of individual photographs. The book would not have happened without you.

A couple of people deserve special mention: Jill Stoddard, then at the Earth Institute at Columbia University, helped outline the idea of the book, pick the chapter authors, and drag people to the meetings where we got the chance to present our ideas; and at W. W. Norton, Leo Wiegman guided us through the process and saw a coherent book long before we figured out what we were doing. We must also give our thanks to Rick Sammon (www.ricksammon.com), who first took our proposal to the publishers.

This book was the outgrowth of a gallery show titled "Photographers' Perspectives on Climate Change." Jessica Bonenfant of the Odonata Dance Project spent many hours helping us through the process of organizing an exhibition, and Christian Paniagua (www.christianpaniagua.com) designed the original poster.

We'd like to thank everyone at Norton who contributed to the final product, especially editorial assistants Lisa Rand and Jennifer Cantelmi; Tom Mayer, who took hold of the editorial reins halfway through; and in particular, Devon Zahn, the production manager. In addition, Jessie Hughes has been amazingly helpful, handling all the contracts, releases, and permissions. We'd like to thank Brittany Morford for editing and stress-management support. Andrew Fountain, Vladimir Romanovsky, Erik Born, Jackie Grebmeier, Sven Haakanson, Robin Bell, and Geoff Brackett made helpful comments on the Arctic chapter. Heather Coiner helped us track down and understand kudzu. We thank Chantal Thompson for her editorial support on the mitigation chapter. Alexandra Allen has our thanks for the countless hours of necessary shuttling of contracts and papers. We also have to thank Barbara

Flotte and Mark Wolfe, who assisted with this project at every stage in too many ways to count. The Carmine Street Jugglers also deserve some blame.

We'd like to thank the following people for generously allowing us to use their images: Katey Walter and Marmian Grimes from the University of Alaska, Scripps Institution of Oceanography, William Chapman, Don Perovich, Bruno Tremblay, Woods Hole Oceanographic Institution, Ken Mankoff, Meyer Steinberg, David Burdick, Roger Angel, Heidi Godfrey, Zafer Kizilkaya (www.imagesandstories .com), Jordan Watson (www.theeighthcontinent.com), and Ashley Cooper (www .globalwarmingimages.net). NASA Public Affairs was extremely helpful in tracking down and acquiring images from their collection.

Finally we must thank everyone at the Earth Institute at Columbia University (particularly Mark Inglis, Mary Tobin, Ken Kostel, Claire Oh, Jennifer Genrich, and Mary-Elena Carr) and NASA. Both organizations have been extremely supportive.

FURTHER READING

Although we've tried to be as comprehensive as space allowed, we have only skimmed the surface of what is known about many important issues. Fortunately, there are plenty of places to find more information, and we list some of the more reliable ones here. Readers should be aware that there are many sources of unreliable information as well, and we caution against assuming a source is correct just because it is highly ranked in a Google or Amazon search.

THE SCIENCE AND HISTORY OF CLIMATE CHANGE

The most comprehensive introduction to the science (and the technical literature that underpins it) is the report from the Intergovernmental Panel on Climate Change. The most up-to-date publication is the *Fourth Assessment Report: Climate Change 2007*, which is available at www.ipcc.ch and in bookstores. It consists of three volumes:

- Working Group I, "The Physical Science Basis" (http://ipcc-wg1.ucar.edu/wg1/wg1-report.html)
- Working Group II, "Impacts, Adaptation and Vulnerability" (www.ipcc-wg2.org)
- Working Group III, "Mitigation of Climate Change" (www.mnp.nl/ipcc/pages_media/AR4-chapters.html)

The Synthesis Report (www.ipcc.ch/ipccreports/ar4-syr.htm) is a summary of all three Working Groups and is probably the gentlest introduction to the material. The Arctic Climate Impact Assessment (www.acia.uaf.edu) is at a similar level, but is focused more on the Arctic.

A more pedagogical treatment can be found in these excellent textbooks:

- *Earth's Climate: Past and Future* by William F. Ruddiman (Freeman, 2001)
- *Global Warming: Understanding the Forecast* by David Archer (Wiley-Blackwell, 2006)

Climate change also has been handled well in more popular formats:

- *Field Notes from a Catastrophe: Man, Nature, and Climate Change* by Elizabeth Kolbert (Bloomsbury USA, 2006)
- *Discovery of Global Warming* by Spencer R. Weart (Harvard University Press, 2004, and www.aip.org/history/climate/index.html)
- *The Winds of Change: Climate, Weather, and the Destruction of Civilizations* by Eugene Linden (Simon & Schuster, 2007)
- *An Inconvenient Truth* by Al Gore (Rodale Books, 2006; DVD from Paramount Pictures)
- *The North Pole Was Here* by Andrew C. Revkin (Kingfisher, 2007)
- *Earth Under Fire: How Global Warming Is Changing the World* by Gary Braasch (University of California Press, 2007)

More specialized books, such as those on hurricanes or specific scientists, are also interesting:

- *Divine Wind: The History and Science of Hurricanes* by Kerry Emanuel (Oxford University Press, 2005)
- *Storm World: Hurricanes, Politics, and the Battle over Global Warming* by Chris Mooney (Harcourt, 2007)
- *Thin Ice: Unlocking the Secrets of Climate in the World's Highest Mountains* by Mark Bowen (Holt, 2006)
- *Our Changing Planet: The View from Space* edited by Michal D. King et al. (Cambridge University Press, 2007)

DEALING WITH CLIMATE CHANGE

Balanced descriptions of policies for coping with and preventing further climate change, or how climate change fits in with other environmental and economic problems facing societies, are rare, unfortunately, but a few titles stand out:

- *Collapse: How Societies Choose to Fail or Succeed* by Jared Diamond (Penguin, 2005)
- *Fixing Climate: What Past Climate Changes Reveal about the Current Threat— and How to Counter It* by Wallace S. Broecker and Robert Kunzig (Hill and Wang, 2008)
- *The End of Poverty: Economic Possibilities for Our Time* by Jeffrey D. Sachs (Penguin, 2005)

- *Confronting Climate Change: Avoiding the Unmanageable and Managing the Unavoidable* by the UN Foundation and Sigma Xi (www.sigmaxi.org/about/news/UNSEGReport.shtml)
- *Hell and High Water: Global Warming—the Solution and the Politics—and What We Should Do* by Joseph Romm (William Morrow, 2007)
- *Earth: The Sequel* by Fred Krupp and Miriam Horn (W. W. Norton, 2008)

ONLINE RESOURCES

Many excellent resources are available on the Internet:

- Dot Earth (http://dotearth.blogs.nytimes.com) Reporter Andy C. Revkin examines efforts to balance human affairs with the planet's limits.
- RealClimate (www.realclimate.org) A blog about climate science by climate scientists.
- New Scientist Environment (http://environment.newscientist.com/home.ns) Environmental news.
- BBC Climate Change (www.bbc.co.uk/climate/)
- Nature Reports Climate Change (www.nature.com/climate/)
- Business of Green blog at the International Herald Tribune (http://blogs.iht.com/tribtalk/business/green/index.php)
- Grist (www.grist.org) Online magazine covering climate and other environmental issues.
- ClimatePolicy (www.climatepolicy.org) Policy discussions from the American Meteorological Society.
- Cleantech Blog (www.cleantechblog.com) Discussions about new technologies.
- *Forecast Earth* at the Weather Channel (http://climate.weather.com)
- NASA Earth Observatory (http://earthobservatory.nasa.gov)
- *Wall Street Journal* Environmental Capital blog (http://blogs.wsj.com/environmentalcapital)
- Encyclopedia of Earth (www.eoearth.org) Expert articles about the Earth, its natural environments, and their interaction with society

GETTING INVOLVED

If you want to calculate your carbon footprint, find out how to contact your representatives, or learn how you can make a difference, there are plenty of places that can help:

- Climate Central (www.climatecentral.org)
- Pew Center on Global Climate Change (www.pewclimate.org)
- Rocky Mountain Institute (www.rmi.org) Focuses on improving energy efficiency.
- Energy Star (www.energystar.gov) A program designed to help you choose the most energy-efficient appliances.
- 350 (www.350.org)
- We Campaign (www.wecansolveit.org)
- 1 Sky (www.1sky.org)
- Footprint Calculator (www.earthday.net/footprint/index.html)

CONTRIBUTOR BIOGRAPHIES

© GARY BRAASCH

Gary Braasch is a conservation photographer and photojournalist who since 1980 has documented natural history and environmental issues, including global climate change. His book *Earth Under Fire: How Global Warming Is Changing the World* (University of California Press, 2007) has been named one of the fifty best environmental books and videos by *Vanity Fair* magazine. He has produced photographic assignments for major magazines, including *National Geographic, Scientific American, Smithsonian,* and *Life.* Mr. Braasch has been named Outstanding Nature Photographer by the North American Nature Photography Association and is the 2006 recipient of the Ansel Adams Award for Conservation Photography by the Sierra Club. He is a founding Fellow of the International League of Conservation Photographers.

© EMANUELE DI LORENZO

Kim Cobb is an assistant professor in the School of Earth and Atmospheric Sciences at the Georgia Institute of Technology. Her research uses geochemical signals locked in ancient carbonates, specifically corals and cave stalagmites, to reconstruct tropical Pacific temperature and rainfall patterns over the last decades to millennia. She received a B.A. from Yale University in 1996, majoring in biology and geology. After earning a Ph.D. in oceanography in 2002 from the Scripps Institute of Oceanography in La Jolla, California, she spent two years at Caltech in the Department of Geological and Planetary Sciences as a postdoctoral fellow before joining the faculty at Georgia Tech in 2004.

© PETER AND YOKO DEMENOCAL

Peter deMenocal is a professor in the Department of Earth and Environmental Sciences at Columbia University. At the Lamont-Doherty Earth Observatory at Columbia University he uses stable isotopic and other geochemical analyses of marine sediments to understand how and why past climates have changed. Current areas of research include: stability of warm climate periods, African monsoonal climate, ancient cultural responses to rapid climate change, and the role of climate change in evolution of early human ancestors. He was awarded the Lenfest Columbia Distinguished Faculty Award in 2008 and is editor in chief of the scientific journal *Earth and Planetary Science Letters.* He has a B.S. in geology from St. Lawrence University, an M.S. in oceanography from the University of Rhode Island Graduate School of Oceanography, and a Ph.D. in geology from Columbia University.

© JOSHUA WOLFE

David Leonard Downie taught courses in environmental politics at Columbia University from 1994 to 2008. While at Columbia he also served as director of Environmental Policy Studies at the School of International and Public Affairs, associate director of the Graduate Program in Climate and Society, and director of the Global Roundtable on Climate Change. He is currently director of environmental studies and associate professor of political science at Fairfield University. Dr. Downie's research focuses on the creation, content, and implementation of international environmental policy. The author of numerous articles and scholarly publications on a variety of topics, his most recent works include co-authoring *Climate Change: A Reference Handbook* and the fourth edition of *Global Environmental Politics*, and editing *The Global Environment: Institutions, Law, and Policy.*

© PETER ESSICK

Peter Essick grew up in Burbank, California, and learned the basics of photography from his father. In 1985, Peter enrolled in the photojournalism program at the University of Missouri. That year he was selected as a photo intern at the National Geographic Society. After earning his graduate degree in photojournalism, Peter began working as a freelance photojournalist. His byline has been in *National Geographic* magazine more than thirty times for stories from around the world, as well as in many other international magazines. He has traveled to all seven continents and all fifty U.S. states in search of compelling pictures. In recent years, he has specialized in stories about nature and the environment. He has illustrated stories on global warming, the carbon cycle, the global freshwater crisis, and nuclear waste.

© LINDA HALL

Tim Hall is an applied physicist who has worked in several areas of climatology. He received a Ph.D. in physics from Cornell University in 1991. Following work in France and Australia, in 1997 he joined the staff of NASA's Goddard Institute for Space Studies in New York City, where he is presently a senior scientist. He is an adjunct professor at Columbia University. Tim's current research addresses the transport of pollutants by currents in the atmosphere and ocean. One application is the uptake by the ocean of industrial carbon dioxide. He also studies tropical cyclones and their relationship with climate. Tim lives with his wife and three children in Cold Spring, New York.

© JOHN KLEINER

Elizabeth Kolbert has been a staff writer for *The New Yorker* since 1999. Her series on global warming, "The Climate of Man," appeared in *The New Yorker* in the spring of 2005. It won a National Magazine Award and was extended into a book, *Field Notes from a Catastrophe*, which was published in 2006. Prior to joining the staff of *The New Yorker*, she was a political reporter for *The New York Times.* She lives in Williamstown, Massachusetts, with her husband and three sons.

© JOHANNES LOSCHNIGG

Johannes Loschnigg was the staff director for the Subcommittee on Space and Aeronautics on the House of Representatives Committee on Science until early 2007, where he previously served as a professional staff member. Johannes first came to Capitol Hill as an American Association for the Advancement of Science congressional science and technology policy fellow in 2002, working in the office of Senator Lieberman. From 1998 to 2002, Dr. Loschnigg was affiliated with the University of Hawaii, where he initially worked as postdoctoral fellow and later became a visiting researcher. In addition, he held positions as graduate research and scientific assistant at the University of Colorado at Boulder, the NASA Ames Research Center, the Department of Physics at the University of Freiburg in Germany, and the Department of Physics at the University of Wisconsin–Madison. Johannes holds a B.A. in both physics and international relations from the University of Wisconsin–Madison, and both an M.S. and a Ph.D. in astrophysical, planetary, and atmospheric sciences from the University of Colorado at Boulder.

© SARA TJOSSEM

Shahid Naeem is professor of ecology, chair of the Department of Ecology, Evolution, and Environmental Biology, and director of science programs at the Center for Environmental Research and Conservation at Columbia University. He works on the ecosystem consequences of biodiversity loss, habitat fragmentation, invasive species, emerging disease, and overharvesting. He received a B.A. and Ph.D. from the University of California, Berkeley, was a Michigan Fellow at the University of Michigan and a postdoctoral researcher at Imperial College London and the University of Copenhagen, and has served on the faculty of the University of Washington in Seattle and the University of Minnesota, Twin Cities. His research is theoretical as well as empirical, and he is currently conducting fieldwork in Inner Mongolia, China; at the Millennium Villages in Africa; and at Black Rock Forest in the Hudson Valley of New York. He has served as cochair on the UN Millennium Assessment's Biodiversity Synthesis Report and is a Fellow of the Aldo Leopold Leadership Program and the American Society for the Advancement of Science.

© HANNAH BELITZ

Naomi Oreskes (Ph.D., Stanford, 1990), professor of history and science studies at the University of California, San Diego, studies the historical development of scientific knowledge, methods, and practices in the Earth and environmental sciences. Her 2004 essay "The Scientific Consensus on Climate Change" (*Science* 306:1686) led to op-ed pieces in *The Washington Post*, the *San Francisco Chronicle*, and the *Los Angeles Times*, and has been widely cited in *The New Yorker*, *USA Today*, the Royal Society's publication "A guide to facts and fictions about climate change," and in the Academy Award–winning film *An Inconvenient Truth*. She is the author of *Science on a Mission: American Oceanography in the Cold War and Beyond*, forthcoming from the University of Chicago Press, and *Fighting Facts*, forthcoming from Bloomsbury.

© BRUCE GILBERT

Stephanie Pfirman is Alena Wels Hirschorn '58 and Martin Hirschorn Professor and chair of the Department of Environmental Science at Barnard College, Columbia University. She received a Ph.D. in marine geology and geophysics from the Massachusetts Institute of Technology/Woods Hole Oceanographic Institution Joint Program in Oceanography and Oceanographic Engineering, and a B.A. in geology from Colgate University. Throughout her career, Dr. Pfirman has been engaged in Arctic environmental research (including sea ice dynamics and glaciers), undergraduate education, and public outreach. She is president of the Council of Environmental Deans and Directors, a member of the National Academy of Science's Polar Research Board, and has chaired the National Science Foundation's Advisory Committee to the Office of Polar Programs and its Advisory Committee for Environmental Research and Education. Prior to joining Barnard, Dr. Pfirman was a senior scientist at the Environmental Defense Fund, where she worked with the American Museum of Natural History to develop the award-winning exhibition "Global Warming: Understanding the Forecast." She was also research scientist at the University of Kiel and IFM-GEOMAR in Germany, and staff scientist for the U.S. House of Representatives Committee on Science.

© JOSHUA WOLFE

Anastasia Romanou is a physical oceanographer and a climate modeler, keenly interested in anything that's at or below sea level in the climate system. Her research interests include ocean biogeochemical cycles and their effects on carbon and climate change. She graduated from the Department of Physics at the University of Athens, received a Ph.D. from Florida State University, and did her postdoctoral work at the Rosenstiel School of Marine and Atmospheric Science at the University of Miami, Los Alamos National Laboratory, and New York University's Courant Institute of Mathematical Sciences. She is now at Columbia University and NASA's Goddard Institute for Space Studies. When she is not delving into science, she and her husband George are raising their two children, Vasily and Marina.

© BRUCE GILBERT

Jeffrey D. Sachs is the director of the Earth Institute, Quetelet Professor of Sustainable Development, and professor of health policy and management at Columbia University. He is also special advisor to United Nations Secretary-General Ban Ki-moon. From 2002 to 2006, he was director of the UN Millennium Project and special advisor to United Nations Secretary-General Kofi Annan on the Millennium Development Goals, the internationally agreed goals to reduce extreme poverty, disease, and hunger by the year 2015. Sachs is also president and cofounder of Millennium Promise Alliance, a nonprofit organization aimed at ending extreme global poverty. In 2004 and 2005 he was named among the 100 most influential leaders in the world by *Time* magazine. He is author of hundreds of scholarly articles and many books, including the *New York Times* bestsellers *Common Wealth* (Penguin, 2008) and *The End of Poverty* (Penguin, 2005). Prior to joining Columbia, he spent over twenty years at Harvard University, most recently as director of the Center for International Development. A native of Detroit, Michigan, Sachs received his B.A., M.A., and Ph.D. degrees at Harvard University.

© JOSHUA WOLFE

Gavin Schmidt is a climate modeler at the NASA Goddard Institute for Space Studies in New York. He received a B.A. (Hons) in mathematics from Oxford University, a Ph.D. in applied mathematics from University College London, and was a National Oceanic and Atmospheric Administration Postdoctoral Fellow in Climate and Global Change Research. He was cited by *Scientific American* as one of the 50 Research Leaders of 2004, and is an advisor to *Popular Mechanics.* He has worked on education and outreach with the American Museum of Natural History, the Collège de France, The New York Academy of Sciences, and *The Daily Show with Jon Stewart.* He is a cofounder of RealClimate.org, where he is a contributing editor.

© CHUN-CHIEH WU

Adam Sobel grew up in New York City and became interested in science partly through childhood visits to the Museum of Natural History and Hayden Planetarium. He attended Wesleyan University, where he double-majored in physics and music, playing jazz trumpet. After graduation, he spent a period trying to be a musician and working as a sound engineer doing audio postproduction for radio and television commercials. He attended the Massachusetts Institute of Technology as a graduate student in meteorology and received a Ph.D. in 1998. He spent two years as a postdoctoral research scientist at the University of Washington in Seattle, then joined the faculty of Columbia University in 2000. He is jointly appointed between the Department of Applied Physics and Applied Mathematics and the Department of Earth and Environmental Sciences. He tries to understand the mechanisms that determine the weather and climate of the atmosphere, especially in the tropics. He is married and has two boys.

© WHITNEY RAE FRUIN

Lyndon Valicenti worked on the staff of the Earth Institute at Columbia University's Global Roundtable on Climate Change. She holds an M.P.A. in environmental science and policy from Columbia University and a B.S. in aquatic biology from the University of California, Santa Barbara. Before graduate school, Lyndon studied the impacts of climate change on shifting zooplankton communities along the Antarctic Peninsula and on phytoplankton in an Arctic lake on Alaska's North Slope. She currently works for the City of Chicago Department of Environment.

© MARK WOLFE

Joshua Wolfe is a freelance documentary and commercial photographer based in New York City. The founder and president of GHG Photos, a collective of photographers who focus on issues related to climate and environment, he is represented by the Henry Gregg Gallery, where he also curates photography shows. Josh studied photography at the Corcoran College of Art, George Washington University, and Pratt Institute. His latest project can be seen at www.theglacierproject.com.

© CHANTAL THOMPSON

Frank Zeman is the director of the Center for Metropolitan Sustainability at the New York Institute of Technology. He is a professional engineer who spent several years in the international mining industry before turning to research in sustainability. Publications include work on the capture of carbon dioxide from ambient air, synthetic fuel production, and sustainable cement production. He is the lead author of a report on reduced carbon dioxide emission kilns for the Cement Sustainability Initiative of the World Business Council on Sustainable Development. He earned a bachelors degree with honors at Queen's University in Canada in 1993, followed by a masters in civil engineering at Imperial College London in 1999. His doctorate was obtained in 2006 from the Department of Earth and Environmental Engineering at Columbia University.

INDEX

Page numbers in *italics* refer to illustrations.

atmosphere (*continued*)

    weather patterns in, 21–22

    see also air pollution; greenhouse gases; stratosphere

Austin, University of Texas at, Visualization Lab of, 179, *190*

Australia, 235, *237*, 253, 272, 274

Axel Heiberg Island, 2

B

Bali climate talks, *251*, 272–73, *273*, 274

Barents Sea, 63, *65*

Barker, James, 66

Barrow, Alaska, 55

Basel, Switzerland, 20

BASF Corporation, 229

Beaufort Sea, 63, 157

beetles, 36, 109, *123*, 124

Bering Sea, 63, *65*, *174*

Bering Strait, *47*, *62*

Berra, Yogi, 195

Bierce, Ambrose, 73

biodiversity, *82*, 114–19, *120*, 124, 125–31, 206–7

biofuel, 227

biomass growth, *260*

biomes, 116–17

biosphere, 1, 113–31

birds, 63, 65, 66, *116*, 121, 123, 222

Bismark, Otto von, 251

black carbon (soot), 148, 149, 228

Blair, Tony, 251

blister-rust fungal disease, 36

Boehlert, Sherwood, 278

Bohr, Niels, 195

Bonneville Dam, *217*

borehole temperatures, 25

boundary-value problem,climate as a, 197

British Columbia, 266, 267

Broecker, Wallace S., 179, *192*

bryophytes (mosses), 178–79, *188*

building codes, 268–69

Bush administration, 274, 275

C

CAFE standards, 274

California, 84, 207–8, 214, *215*, *225*, 274

Callendar, Guy, 154

Canada, 54, 57, 60, 67, 109, 127, 226, 253, 266, 267

Canadian Archipelago, 48, 51

cap-and-trade systems, 263–65

Cape Hatteras, North Carolina, 35–36, *41*

Capitol power plant, District of Columbia, *257*

carbon, 48, 69, 78, 89, 144, 145, *169*, 216

    black (soot), 148, 149, 228

    capture and sequestration (CCS) , 214, 224–27, *225*, *231*, *260*

    small and mobile sources of, 227–28

    taxes, 263, 265–67

carbonate, 79–80, 82–83

carbon cycle, 143–48, *144*, 165, 198, 226, 232, 253

carbon dioxide, 6, 70, 80, 114, 125, 128, 131, 136, 137, 139, 164–65, 205, 229, 231, *254*, 261

    cap-and-trade systems for, 263–65

    climate change and, 140, 141, 153, 154

    cost of reducing emissions of, 261–62, 263

    levels of, 137–38, *138*, 145–48, 198–99

    oceans' absorption of, 75, 79–80

    reducing emissions of, 214, 224–29, *225*, 251, 252, 255–56, 275

    released by melting permafrost, 62

    sequestering of, *231*, 232–33

carbon footprint, 269

carbon sinks, 147, 261, 267

Caribbean Ocean, 81

caribou, 58

Carnot, Sadi, 218

Carroll, Lewis, 10

cars, 268, 270

Cassini, Giovanni, *159*

caterpillars, *130*

CFCs (chlorofluorocarbons):

    banning of, *5*, 6, 139, 229

    chemical components of, 4

    as greenhouse gases, 6, 136, 139

    ozone depletion caused by, 4–5, *5*, 139, 229, 254

Chad, Lake, *102*, 103, *104*

Chicago, Illinois, 35, *37*, *268*

China, *140*, *145*, 149, 220, 228, 255, 259, 263, 273, 275

chlorine, 4–5, *5*, 6

Christmas Island (Pacific Ocean), 90, 92, 93

cities, 159–60

    climate change adaptation in, 235–36, *237*, *238*, *240*, *242*, *244*, *246*, *248*

    coastal, *85*, 86, *87*

    urban heat island effect and, 21, 22

Clean Air Act (U.S.), 224, 263, 274
climate:
    biodiversity and, 117, 131
    definition of, 1
    drivers of, 135–52, *135*
    feedbacks, 11
    influence of, 2
    observations of, 158–62, *158, 159, 160, 161*
    range of configurations of, 130
Climate Action Partnership, 258
climate change:
    adaptation to, 2, 3, 208–9, 235–36, *237, 238, 240, 242, 244, 246, 248,* 258–59
    biodiversity and, 118–19
    business-as-usual approach to, 3, 9, 11, 199, 209, 255, 258
    Congress and, 276–78
    consensus to act against, 256–58
    corporate community's awareness of, 257–58
    cost of combating, *see* economics
    definition of, 2–3, 95, 231
    diagnosis of, 3, 133–210
    forecasts, predictions, and projections for, 195–98
    glossary of terms relevant to, 10–15
    historical, 2
    impacts on ecosystems, 119–25, 168–69,
    individual action and, 269–72
    international cooperation on, 272–75
    interpreting scenarios for, 199–200, *200*
    mitigation of, *see* mitigation strategies
    mortality due to, 35
    natural vs. human-caused (anthropogenic), 8, 90, 93, 97, 151–52, 153
    politics and policy decisions and, 9, 209, 252–56, 262–69
    possible cures for, 3, 9, 211–78
    prognosis for, 195–209
    proxy records of, 27–30, 163–68, *163, 165, 166, 168*
    public interest in, 8, 280
    reporting on, 70–72
    "scientization" of political debate over, 8
    symptoms of, 3, 17–132
    timescale and, 279
    unknowns regarding, 209
    see also specific causes and effects
climate models, 14, 103–4, 107, 109, 141–42, 150, 153, 169–74, *170, 171, 172,* 176, *196,* 197–99

climate sensitivity, 141
    see also feedback
climatologists, 157–58, 174, 178–94
climatology, 1, 157–77
clouds, 47, 142–43, *143,* 148–49, 160–61
coal, *140, 145, 146,* 216, 224, 225, *254, 257,* 263, 264
coastal development, *85*
coastal erosion, 35–36, *40,* 62–63, *62,* 86
cod, 85
cold snaps, 109, 207
Coleridge, Samuel, 45
Colorado River, 213
    see also Hoover Dam; Mead, Lake; Powell, Lake
Columbia Glacier, Alaska, 54, 178, *180*
Columbia University, Lamont-Doherty Earth Observatory at, *28, 192–93*
compact fluorescent light (CFL) bulbs, *218,* 219, 251, 270, 274
congestion charges, 268
Congress, U.S., 274, 275, 276–78
continental drift, 2, 7, 135, 154
Cook, Ed, *28*
corals, 27–28, 29, 80–83, *81, 82,* 90–93, *91, 92,* 118, 124–25, 127, *127,* 163, 207, 226
Cordillera Blanca (White Mountain Range), Peru, 32–34, *32*
corn ethanol, 215, 227
corporate action on climate change, 257–58
Cretaceous period, 2, 119
Crutzen, Paul, 5
cyclones, *see* tropical cyclones

D

D'Arrigo, Rosanne, *28*
Day After Tomorrow, The, 205
Delta Works project, Netherlands, 235
Denbies Wine Estate, UK, *208*
desalination plants, 235, *237*
deserts, 69, 117, 165, 167
developed nations, 253
Devil's Dictionary, The (Bierce), 73
diatoms, 207
dinosaurs, 2, 135
Domesday Book, *208*
Doncaster, UK, *108*
Dorale, Jeff, *163*
Dow Chemicals, 274

oceans (*continued*)
    carbon storage in, 144, 147, 226
    chemical changes in, 78–80
    cold tongue in East Pacific, 74–75, *74*
    density, 74
    estuaries and coastal regions of, 85–88, *85, 87, 88*
    food chain in, 80, 85
    greenhouse gases absorbed by, 75, 78, 79–80
    heat transport by, 46
    increasing acidity of, 78, 79–80, 82, 207, 226
    salinity, 74, 78–79
    sediment cores taken from, 165–66, *166, 174*
    temperature, 22–25, *25,* 93, 99, 103
    thermal inertia of, 23, 75, 150, 209, 253
    thermohaline circulation of, 50, 74–75, 89, 174
    topography of, 76–77
    as unknown territory, 88–89
    "weather" of, 197
    see also corals; fisheries; sea levels; *specific oceans*
oil, 215–16, 226, 275
Oregon, 35, *40*
organic pollutants, 66–67
oxy-combustion, *see* carbon, capture and sequestration
oxygen isotopes, 163–64, *163,* 165–66
ozone (tropospheric), 136, 140, 229
ozone depletion, 3, 4–7, 209, 253–54, 280
    CFCs and, 4–6, *5,* 139, 229
    connection to climate change, 6, 140
    international cooperation and, 6–7
    in northern latitudes, 66
    model failures, 6
    politics and, 6
ozone layer (stratospheric), 231
    expected recovery of, *5,* 6
    location of, 4
    ultraviolet radiation blocked by, 3, 113
    see also ozone depletion

P

Pacala, Stephen, 233
Pacific Ocean, *25, 77,* 96, 105
    climate change in, 93
    changing currents in, 205
    fisheries of, 83–84
    long-term temperature variability of, 29
    plastic ducks as tracer in, *73, 73,* 75
pack rat middens, 167, *168*

Pagati, Jeff, 178, *182*
paleobiology, 119
paleoclimate research, 27–30, 90–93, *91, 92,* 163–68, *163, 165, 168,* 204
Palmyra Island, 90, *92,* 93
Papua New Guinea, 274
Paris Observatory, 159, *159*
PCBs, 66–67
penguins, 121, *121,* 125, 169, 178, *186*
permafrost, 52
    description of, 46, 57
    greenhouse gases and, 48, 62, 69, 147–48, 205
    melting of, 46, 57–63, *57, 58, 59, 62,* 147–48
    projected shifting of, 201
Perth, Australia, 235, *237*
Peru, 32–34, *32*
petroleum, *see* oil
phenology, 36, 118, 119
photosynthesis, 144, 216, 232
photovoltaic (PV) panels, 213–14, 220–21, 223, 234
Pinatubo, Mount, 140, 149, 172, *173*
    see also volcanoes
pine bark beetle, *see* beetles
pipelines, 60
plankton, 63, *65, 79,* 80, 82–83, *84,* 114, 131, 169, 216, 232
plants, 144
    in Arctic, 47, 48, 56, 58, 61
    oxygen-carbon dioxide cycle and, 114
    projected changes in, 36, *42,* 206–7
plate tectonics, 7
Pliocene era, 34
polar amplification, 46, 142, 152
polar bears, *51,* 52, 66, 125, 157
polar regions, 1, 2, 3, 113, 119, 153
    see also Antarctica; Arctic
politicization of science, 8
politics:
    Arctic territories and, 67
    climate change and, 252–56
    ozone depletion and, 6
pollen, 167, *168*
pollinators, 123
polyps (coral), 80–81
Powell, Lake, 35, *38, 106*
power generation, 213–17, 218, 220–23
    base load plants, 220

see also fossil fuels; geothermal energy; hydropower; nuclear energy; solar energy

power grids, 223

predictions, 196

    see also forecasts; projections

Prem-Air "smog eating" system, 229

projections, 196–97

proxy climate change records, 27–30

Q

Quelccaya Ice Cap, Peru, 32–34, *32, 181*

R

radiation:

    definition of, 14

    infrared (longwave), 14, 113, 136, 137, 143

    measuring of changes in, 25–27, *26*

    solar (shortwave), 14, 113, 136, 139–40,

    spectra from Earth, 113–14

radiative forcing, 139–40, 141, 148

radiolaria, 207

rain, 27, 35, 55, 56, 74, *95,* 99, 107, 111, 131, 142, 160, 213, 260

    projected changes in, 202–3, *202*

rainforests, 116, 206

reflectivity, see albedo

Regional Greenhouse Gas Initiative, 267

regulations, 263, 267–68

reindeer, 56–57

Revelle, Roger, 79, 135, 138

reverse osmosis, *237*

Reykjavik, *222*

"Rime of the Ancient Mariner, The" (Coleridge), 45

Rio Earth Summit (1992), xi, 209, 272

rivers, 203, 207

Rotterdam, Netherlands, 235, *238*

Rowland, Sherwood, 5, 280

Russell, Gary, *171*

Russia, 54, 56, 60, 62, 67, 253, 274

    see also Siberia

S

Saami people, 56

Sahara, 119, 166–67, 172–73

Sahel, the, 99, 102–4, *102, 104,* 202

salmon, 84

Saudi Arabia, 275

savannas, 116, 120, 206

Scandinavia, 30, 57

    see also specific countries

science:

    abuse and distortion of, 8–9, 153–54

    consensus in, 153–55

    policymaking and, 256

    proof and, 151–52

scientific method, 7–9, 196

scientists:

    educational role of, 9

    technical use of language by, 10

    see also climatologists

Scotland, 27

sea ice, see Antarctic; Arctic sea ice

sea levels, *85,* 209, 235, 236, 256, 260

    coastal erosion and, 63

    in Eemian interglacial period, 204

    effects of rise of, *203*

    eustatic (global) rise in, 77–78, 141

    glacier melting and, 54, *68*

    influences on, 76–78

    in Pliocene era, 34

    projected rise of, 203–4, *203*

    rise since the last ice age, 2, 57

sea lions, 844

seals, *48,* 52, 65, 121, 123, 125

seasonal cycle, 1

Seneca, 113

Shakespeare, William, 95

shipping, 67

    melting of Arctic sea ice as benefit to, 47, 51–52

    tracks, *134*

    see also transportation

Shishmaref, Alaska, *62,* 70

Siberia, 45, 48, 51, 57, 58, 60, 67, 201, *206*

Sleipner, Norway (carbon storage project), 226

Smits, Arie and Marianne, *248*

"smog eating" radiator coating, 229

snow, snowfall, snowmelt:

    in Antarctica, 69

    black carbon effect on, 149

    glaciers and, 30, 31, 54, 55

Snowball Earth, 2

Socolow, Robert, 233

soil erosion, 231

solar cells, *215*

U

ultraviolet (UV) radiation, 3, 4, 66, 113, 136, 137, 229
United Nations, 259, 277
United Nations Framework Convention on Climate Change (UNFCCC), 272
units (conversions from metric), 15
urban heat island effect, 21, 22
urbanization, 207
urban planning, 268

V

Venice, 86, 204, 236, *244, 246*
Villon, François, 19
vineyards, 118, 207–8, *208*
Virginia bluebell, 36, *42*
volcanoes, 30, 140, 149, 172, *173,* 197, 198, 231
Volvo, 229

W

Wal-Mart, 251, 274
Walter, Katey, *63*
Washburn, Mount, Wyoming, 36, *43*
Washington, D.C., *257*
water, 164, 213
    fresh, 74
    for irrigation and damming of rivers, 78–79, 103, 128
    stored by glaciers, 34, 69
    see also hydropower; oceans; rain; rivers
water vapor, 136, 141–42
    see also humidity
Watt-Cloutier, Sheila, 45
weather:
    chaotic nature of, 197
    climate and, 1, 22, 95–96, 197

extremes of, 95–111
    variability of, 21–22
weather satellites, 25–27, *26,* 53, 97, *135, 149, 157,* 162, 179
weather stations, 20–21, *20,* 22, 158, 159–60
Wegener, Alfred, 7, 154
Western Climate Initiative, 267
West Pacific Warm Pool, 76
wetlands, 58, 60, 87–88, 138, 207, 216
Weyburn, Saskatchewan (carbon storage project), 226
whales, 63, 65–66, 84, 115, 121, 123, 125
whitebark pine, 36, *43*
White Mountain Range (Cordillera Blanca), Peru, 32–34, *32*
Wikipedia, 227
Wiles, Greg, 178, *180*
wind power, 220, 221–22, *221,* 223, 270, *270*
winds, 2, 69, 84
    Antarctic ozone hole and, 6
    changes in, 205
    changes in patterns of, 27
    impact on Arctic sea ice, *47,* 48, 51
    ocean currents and, 73–74
    westerly, 27, 55, 73, 86, 172
wine, *see* vineyards
World Bank, 252, 262
World Health Organization (WHO), 128

Y

"Year without a Summer" (1816), 30
Yellowstone National Park, 36, *43*
Yucatán, Mexico, 30

Z

Zagorodnov, Victor, *181*
zero-waste society, 234
Zierikzee, Netherlands, 235, *240*

Copyright © 2009 by Gavin Schmidt and Joshua Wolfe

Foreword copyright © 2009 by Jeffery D. Sachs

The photographs in this book are copyright the individual photographers and institutions listed on the captions.

All rights reserved

Printed in China

First Edition

For information about permission to reproduce selections from this book, write to Permissions, W. W. Norton & Company, Inc., 500 Fifth Avenue, New York, NY 10110

For information about special discounts for bulk purchases, please contact W. W. Norton Special Sales at specialsales@wwnorton.com or 800-233-4830

Manufacturing by Midas Printing

Book design and composition by Laura Lindgren

Production manager: Devon Zahn

Library of Congress Cataloging-in-Publication Data

Schmidt, Gavin.

Climate change: picturing the science / Gavin Schmidt and Joshua Wolfe.

p. cm.

Includes bibliographical references and index.

ISBN 978-0-393-33125-7 (pbk.)

1. Climatic changes. 2. Climatology. I. Wolfe, Joshua, 1983–   II. Title.

QC981.8.C5S3556 2009

551.6—dc22   2008035134

W. W. Norton & Company, Inc., 500 Fifth Avenue, New York, N.Y. 10110

www.wwnorton.com

W. W. Norton & Company Ltd., Castle House, 75/76 Wells Street, London W1T 3QT

1 2 3 4 5 6 7 8 9 0